NITRATE IN LEAFY VEGETABLES

TOXICITY AND SAFETY MEASURES

NITRATE IN LEAFY VEGETABLES

TOXICITY AND SAFETY MEASURES

Editors
Shahid Umar
Naser A. Anjum
Nafees A. Khan

I.K. International Publishing House Pvt. Ltd.

NEW DELHI • BANGALORE

Published by
I.K. International Publishing House Pvt. Ltd.
S-25, Green Park Extension
Uphaar Cinema Market
New Delhi – 110 016 (India)
E-mail: info@ikinternational.com
Website: www.ikbooks.com

ISBN 978-93-82332-21-3

10 9 8 7 6 5 4 3 2 1

Published by Krishan Makhijani for I.K. International Publishing House Pvt. Ltd., S-25, Green Park Extension, Uphaar Cinema Market, New Delhi – 110 016 and Printed by Rekha Printers Pvt. Ltd., Okhla Industrial Area, Phase-II, New Delhi – 110 020.

Foreword

Vegetables have been a significant part of human diet since time immemorial. Besides adding the elegance and attractiveness to a meal, vegetables are abundant source of vital minerals, vitamins and several biologically active compounds essential for maintaining human health. Most importantly, vegetables account for 70-85% of total human nitrate intake nitrate ions—ubiquitous within food and physiological systems (Huarte-Mendicoa et al. 1997; Moorcroft et al. 2001); after ingestion, the ingested nitrate gets converted into nitrate by microflora in the oral civility and in the gastrointestinal tract. This results in increased oxidation of hemoglobin to methemoglobin, leading to methaemiglobinnemia. Simultaneously, increased production of free oxide radical and free radical nitrate oxide occurs. These radicles predispose person for carcinogenic and other effects. The other effects observed were increased infant mortality, abortions, birth defects, recurrent diarrhea, recurrent stomatitis, early onset of hypertensions, histopathological changes in cardiac muscles, alveoli of lungs and adrenal glands, recurrent respiratory tract infection in children, hypothyroidism and diabetes. Recent ongoing studies indicated that nitrate ingestion adversely affects the immune system of the body as well.

Among vegetables, mainly the green leafy vegetables (such as lettuce, spinach, beets, radishes, celery, etc.) contain the highest levels of nitrates (Prasad and Chetty 2008). Moreover, nitrate contamination in vegetables occurs when crops absorb more than they require for their sustainable growth. Hence, nitrate content is considered an important quality characteristic of vegetables; where, excessive-use of nitrogenous fertilizers largely boosts the accumulation of nitrate in leafy vegetables which is wrongly considered by farmers as reasonable insurance against yield loss (Anjana et al. 2007). Therefore, considering severe human health concerns due to high nitrate containing vegetables' ingestion and limited availability of the literature on the nitrate in leafy vegetables, its toxicity and safety measures in humans, the decision of publishing a book focusing mainly on the nitrate in leafy vegetables emphasizing nitrate-mediated toxicity in human health and approaches for safety measures is timely and relevant to the present/future diet scenario.

Editors of *Nitrate in Leafy Vegetables: Toxicity and Safety Measures* have done an admirable job of assembling a wealth of information on the subject contributed by eminent researchers and scientists working in the field of nitrate in soils and plants, and its human health consequences and sustainable solutions.

I am sure that the current comprehensive volume thereby will enlighten readers of various disciplines and at various levels; and should prove useful for advanced students, researchers, faculty of both plant and animal sciences, and environmentalists and policy makers.

(YASH P. ABROL)

(INSA-Honorary Senior Scientist)
Adjunct Professor of Plant Physiology
Department of Botany, Faculty of Science,
Hamdrad University (Jamia Hamdard)
New Delhi – 110 062, India.

Formerly Head: Division of Plant Physiology
Indian Agricultural Research Institute (IARI)
New Delhi –1100 012, India.

President: Indian Nitrogen Group (ING),
Society for Conservation of Nature (SCON),
F-4. A. Block, NASC Complex, New Delhi -1100 012, India

e-mail: ypabrol@gmail.com

REFERENCES

Anjana, Umar, S., Iqbal, M. and Abrol, Y.P. (2007). Are nitrate concentrations in leafy vegetables within safe limits? *Curr. Sci.* **92:** 355-360.

Huarte-Mendicoa, J.C., Astiasaran, I. and Bello, J. (1997). Nitrate and nitrite levels in frozen broccoli: Effect of freezing and cooking. *Food Chem.* **58:** 39-42.

Moorcroft, M.J., Davis, J. and Compton, R.G. (2001). Detection and determination of nitrate and nitrite: a review. *Talanta* **54:** 785-803.

Prasad, S. and Chetty, A.A. (2008). Nitrate-N determination in leafy vegetables: Study of the effects of cooking and freezing. *Food Chem.* **106:** 772-780.

Preface

Nitrate (NO_3^-) and nitrite (NO_2^-) are ubiquitous within environmental, food, industrial and physiological systems, and while our understanding of their role within such matrices has increased, a substantial degree of uncertainty and speculation remains (Moorcroft et al. 2001). When ingested within permissible limit, these ions play significant roles in various physiological functions of human body including their roles in vascular and immune functions. However, excess nitrate and its metabolites are toxic and lead to various ailments in humans (Gupta et al. 2001, 1999a,b; Sharma and Sharma 2012). Nitrates enter human body through sources such as drinking water, food and air. Out of the total human nitrate intake, fruit and vegetables account for 70%, drinking water 21%, and meat and meat products 6%. However, the common nitrate-rich vegetables are lettuce, beetroot, celery, egg plant, beet, banana, strawberry, tomatoes and peas. The leafy vegetables are important source of nitrates in human body (Dich et al. 1996; Hord et al. 2009). In fact, nitrate content is an important quality characteristic of vegetables. The use of excessive nitrogenous fertilizers has been regarded as one of the major reasons leading to accumulation of nitrate in leafy vegetables which is wrongly considered by farmers as reasonable insurance against yield loss. Therefore, vegetable nitrate content is of interest to governments and regulators owing to the possible implications for health and to check that controls on the content are effective (Anjana et al. 2007).

It is worth mentioning here that although there exists a plethora of publications focused on nitrate in soils, plants and their potential impact on human health, the available research reports and findings from different arenas are largely disorganized and not critically cross-linked and/or integrated. Therefore, this book is an effort to provide a common platform for plant scientists and health professionals working with a common aim of sustainable solutions to nitrate-led human and environmental health consequences.

Vegetables constitute a major dietary source of nitrate, contributing to the daily dietary intake for a percentage ranging from 30 to 94%. Although nitrate itself is relatively non-toxic, or even beneficial, the possible harmful effects of nitrate derived compounds on human health arouse public concern and have led to the development of government regulations and monitoring programs aimed at protecting the public health. In this perspective, the *Chapter 1* by Di Gioia et al. focuses on the factors concerning nitrate accumulation in vegetables and their contribution to nitrate intake in the human diet, taking into account the recent EU

Regulation No. 1258/2011 and the Scientific Opinion of the Panel on Contaminants in the Food chain (CONTAM) of the European Food Safety Authority (EFSA). Öztürk et al. in *Chapter 2* present some interesting outcomes from recent experiments performed in Turkey where a large number of producers apply both inorganic as well as organic fertilizers (in a rage of 40–75 kg/da) which is tenfold of the average recommended for the area. In this important perspective, in addition to suggesting important strategies for reducing nitrate accumulation in soil, water and crop, the authors advocate the use of correct nutrient application at optimum rate and appropriate plant phonological growth stage when the crop plants requirement for nutrients is high. The accumulation of NO_3 in the edible parts of varied vegetables depends on various factors, especially the amounts, timing and sources of N fertilizers and cultivars. Therefore, decreasing these types of contaminants is necessary for the promotion of human health. The problem of nitrates is endemic internationally as well as nationally, but so far no compiled data at national level are available. The data available are either of small areas or of scattered zones/ areas. Keeping in view the aforesaid facts, *Chapter 3* (Gupta et al.) overviews: (a) national and international perspectives of nitrate toxicity in humans, (b) major exogenous and endogenous sources of nitrates, (c) total daily intake of nitrate and its kinetics and metabolism, (d) major acute and chronic toxic effects of nitrate ingestion in humans and animals, mechanisms of methemoglobin formation, and (e) major results of nitrate toxicity-based experiments performed in animal and human systems; and critically discusses in detail the major strategies for treatment and prevention of nitrate toxicity. In *Chapter 4*, Malakouti et al. present an overview of sources of nitrate to soil, crop and human health, nitrate metabolism in plants, nitrate contents in edible parts of major vegetables, nitrate risk assessment, and hazardous effect of nitrate on human health and critically discuss major strategies for minimizing the risk of NO_3 contamination. Furthermore, the authors assure that the country's nutritional ranking may improve and the yield and quality of crops, especially taste and shelf-life of vegetables may be upgraded significantly, and hence the problem of malnutrition will be solved by practicing balanced and appropriate/correct fertilization. Organic farming has emerged as an outcome of consumer's reaction against harmful toxins and the desire for healthier life and safe environment. Not only does organic farming supply healthy foods to human beings, but also conserves the environment in which agricultural practices occur. Vegetables occupy an important role in human diet, and are consumed in large quantities. However, vegetables, especially leafy vegetables, could contain some undesired compounds that are hazardous for human health like nitrate, being toxic to humans in higher concentrations. Through case studies performed on spinach and cabbage - two of the most important leafy vegetables in terms of their cultivation area and consumption rate in the world, *Chapter 5* (Sonmez and Citak) discusses important factors in maintaining the nitrate level in these vegetable crops and the effectiveness of the farmyard manure, chicken manure, guano type and blood meal in accumulation of nitrate in vegetables. Soil is the only finite resource to feed the increasing population, and soil fertility is vital to sustain the food productivity of soil. Additionally, fertilizers are required for sustained food production but their widespread use has roused concern about resulting environmental pollution. In this context, hundreds and thousands of tons of chemical and organic fertilizers are produced globally each year for application in agricultural lands; however, it is estimated that only about half of the applied fertilizers goes into the crops. Therefore, the potential economic losses and negative environmental impacts are obvious. Nevertheless, chief among these problems are the excessive concentration of N compounds in the atmosphere (green house gasses), contributions of phosphate and nitrogen compounds to the eutrophication of surface waters, and risk to human health from manure associated pathogens. Keeping in view these aspects, *Chapter 6* (Ahmad et al.) presents a critical discussion on some of these important issues taking into consideration especially the use of chemical-based fertilizers versus organic fertilizer for crop nutrition and the assessment of its potential impact on environmental and threats to human health in detail. A number of concerted efforts have been made to reduce the excessive nitrogen fertilizer use-mediated negative environmental impacts from agronomic to genetic engineering. However, the potential, strategies, namely, the use selection of nitrogen-efficient crops and/or adoption of strategies for enhancing nitrogen-use efficiency may largely work in isolation or combination in reducing ingestion of crop products with high nitrate-mediated anomalies in human. To this end, *Chapter 7* (Umar et al.) overviews the significance of plant nutrients for enhancing crop production, highlights nitrogen use in leafy vegetables production and critically discusses important approaches useful for enhancing nitrogen-use efficiency in crops in general and leafy vegetable crops in particular. Balancing the amount of N needed for optimum plant growth while reducing nitrate accumulation in plants and other N losses from soil remains a major challenge for everyone attempting to understand and improve agricultural nutrient use efficiency. Considering these aspects, *Chapter 8* (Singh et al.) discusses current crop, soil, and N fertilization management strategies and identifies future research needs aimed at reducing nitrate accumulation in plants. In this chapter, the authors advocate the combination of soil management practices and inherent physical, chemical, and biological characteristics of the soil as well as the physiological and environmental factors as the main reasons for increasing nitrate accumulation in varied environmental compartments, food and human health. Nitrogen losses from soil to water and air have become a global concern. Attempts are being made to put forth some of the best management practices to be used for controlling nitrogen loss from the fields or glasshouses.

In conclusion, *Nitrate in Leafy Vegetables: Toxicity and Safety Measures* represents the first comprehensive compilation of the latest science of the dietary nitrate sources, potential human health effects and sustainable remedial strategies. This text provides practical, data-driven resources based upon the totality of the evidence to help the reader understand the basics, treatments, and preventive strategies that are involved in the understanding all about nitrate in plants and humans. The overarching goal of the scholarly chapters is to provide fully referenced information to plant scientists and health professionals; so that they may have a balanced perspective on the value of various preventive and treatment options that are available today as well as in the foreseeable future.

We would like to acknowledge esteemed contributors who responded to our request and contributed their valuable chapters enthusiastically, containing the latest information on the relevant aspects. The editors acknowledge their family members for their inspiration, endurance and moral support during this period.

We are also thankful to the Potash Research Institute of India (PRII), Gurgaon, INDIA; International Potash Institute (IPI), Switzerland; the Foundation for Science & Technology (FCT), Portugal (SFRH/BPD/64690/2009; SFRH/BPD/84671/2012); Council of Scientific & Industrial Research (CSIR), New Delhi, India; Department of Science and Technology (DST), New Delhi, India and University Grants Commissions (UGC), New Delhi, India for providing partial financial assistance to our research work.

Editors

REFERENCES

Anjana, Umar, S., Iqbal, M. and Abrol, Y.P. (2007). Are nitrate concentrations in leafy vegetables within safe limits? *Curr. Sci.* **92:** 355-360.

Dich, J., Jrvinen, R., Knekt, P. and Penttil, P.L. (1996). Dietary intakes of nitrate, nitrite and NDMA in the finish mobile clinic health examination survey. *Food Add. Cont.* **13:** 541-52.

Gupta, S.K., Gupta, R.C., Gupta, A.B., Seth, A.K., Bassin, J.K. and Gupta, A. (1999). Epidemiological evaluation of recurrent stomatitis, nitrates in drinking water and the cytochrome b5 reductase activity. *Am. J. Gastroenterol.* **94:** 1808-1812.

Gupta, S.K., Gupta, R.C., Gupta, A.B., Seth, A.K., Bassin, J.K. and Gupta, A. (1999). Adaptation of the cytochrome b5 reductase activity and methemoglobinemia in areas with high nitrate concentration in drinking water. *Bull. World Health Org.* **77:** 749-753.

Gupta, S.K., Gupta, R.C., Gupta, A.B., Seth, A.K., Bassin, J.K. and Gupta, A. (2001). Recurrent acute respiratory tract infection in areas with high nitrate concentration in drinking water. *Environ. Health Persp.* **108:** 363-66.

Hord, N.G., Tang, Y. and Bryan, N.S. (2009). Food sources of nitrates and nitrites: the physiologic context for potential health benefits. *Am. J. Clin. Nutr.* **90:** 1-10.

Moorcroft, M.J., Davis, J. and Compton, R.G. (2001). Detection and determination of nitrate and nitrite: a review. *Talanta* **54:** 785-803.

Sharma, M.K. and Sharma, H. (2012). Evaluation of the haematological responses to high nitrate exposure in rabbits. *J. Clin. Diagnos. Res.* **6:** 145-149.

Contents

Contributors

A.B. Gupta: *Department of Environmental Engineering, MNIT, Jaipur – 302 017, Rajasthan, India*

Anjana: *IGNOU Regional Centre, I.T.I. Building, Bulepur, Khanna, Distt.-Ludhiana, Punjab – 141401, India*

Francesco Di Gioia: *Dipartimento di Scienze agro-ambientali e territoriali, Università di Bari, via Amendola 165/A, 70126 Bari, Italy*

Harmandeep Singh: *International Plant Nutrition Institute (IPNI), Western India, C-2/301, Lunkad Gold Coast, Viman Nagar, Pune – 411 014, Maharashtra*

Kaushik Majumdar: *IPNI, India and Bangladesh Programs, 354, Sector-21, HUDA, Gurgaon – 122 016, Haryana, India*

M.J. Malakouti: *Soil Science Department, Tarbiat Modares University. Tehran-Iran*

Maria Gonnella: *Istituto di Scienze delle Produzioni Alimentari (CNR), via Amendola 122, 70126 Bari, Italy*

Mitsuhiro Inoue: *Arid Land Research Center, Tottori University, Japan*

Münir Öztürk: *Ege University, Botany Department, Bornova, Izmir, Turkey*

Nafees A. Khan: *Department of Botany, Faculty of Life Sciences, Aligarh Muslim University, Aligarh – 202 002, U.P., India*

Naser A. Anjum: *Centre for Environmental and Marine Studies (CESAM) & Department of Chemistry, University of Aveiro, 3810-193 Aveiro, Portugal*

Pietro Santamaria: *Dipartimento di Scienze agro-ambientali e territoriali, Università di Bari, via Amendola 165/A, 70126 Bari, Italy*

R.C. Gupta: *Department of Physiology, SMS Medical College, Jaipur – 302 004, Rajasthan, India*

S.J. Tabatabaee: *Horticultural Science Department, Tabriz University. Tabriz-Iran*

Sadahiro Yamamoto: *Faculty of Agriculture, Tottori University, Tottori, Japan*

Sahriye Sonmez: *Akdeniz University, Faculty of Agriculture, Soil Science Department, Antalya/Turkey*

Salih Gücel: *Near East University, Environmental Sciences Institute, Nicosia, Cyprus*

Sedat Citak: *Akdeniz University, Faculty of Agriculture, Soil Science Department, Antalya/Turkey*

Serdal Sakçali: *Fatih University, Biology Department, Istanbul,Turkey*

Sh. Ladan: *Soil Science Department, Tarbiat Modares University, Tehran-Iran*

Shahid Umar: *Department of Botany, Faculty of Science, Hamdard University, New Delhi – 110 062, India*

Sunil Gupta: *Consultant Pediatrician & Neonatologist, and, Scientist of Environmental Medicine, Krishna Ram Ayurvigyan Shodh Sansthan, Jaipur – 302015, Rajasthan, India*

T. Satyanarayana: *IPNI, South India, 401, Tejaswini Arcade, Plot No. 78, 10-2-99, 99/2, Opposite More Super Market, West Marredpally, Secunderabad – 500 026, Andhra Pradesh*

Zahoor Ahmad: *Arid Land Research Center, Tottori University, Japan*

1

Contribution of Leafy Vegetables to Dietary Nitrate Intake and Regulations

Francesco Di Gioia[1], Maria Gonnella[2] and Pietro Santamaria[1]

[1]*Dipartimento di Scienze Agro-Ambientali e Territoriali, Università di Bari, via Amendola 165/A, 70126 Bari, Italy*

[2]*Istituto di Scienze delle Produzioni Alimentari (CNR), via Amendola 122, 70126 Bari, Italy*

e-mail: maria.gonnella@ispa.cnr.it

ABSTRACT

Vegetables constitute a major dietary source of nitrate, contributing to the daily dietary intake for a percentage ranging from 30 to 94%. Although nitrate itself is relatively non-toxic, or even beneficial, the possible harmful effects of nitrate derived compounds on human health arouse public concern and have led to the development of government regulations and monitoring programs aimed at protecting the public health. In this perspective, have been set an acceptable daily intake for nitrates and maximum levels of nitrate content for several vegetables, paying particular attention to some leafy vegetables, characterized by a high capacity to accumulate nitrates. This chapter focuses on the main factors concerning nitrate accumulation in vegetables and their contribution to nitrate intake in the human diet, taking into account the recent EU Regulation No. 1258/2011 and the Scientific Opinion of the Panel on Contaminants in the Food chain (CONTAM) of the European Food Safety Authority (EFSA).

Keywords: Leafy vegetables; Diet; Nitrate; Regulations; Toxicity

INTRODUCTION

Nitrate is a naturally occurring form of nitrogen and is an integral part of the nitrogen cycle in the environment. Nitrate is formed from fertilizers, decaying plants, manure and other organic residues. It is found in the air, soil, water, food, and is produced naturally within the human body (Walker 1990; Gangolli et al. 1994; Speijers 1996; Speijers and van den Brandt 2003). It is also used as a food additive, mainly as a preservative and anti-microbial agent (Speijers 1996; Speijers and van den Brandt 2003). It is used in foods such as cheese and cheese products, raw and processed meats, edible casings, processed fish, fish products, spirits and liqueurs.

Due to the important role of nitrogen as plant nutrient and its dynamic in the agro-ecosystem, growers often tend to over-fertilize their crops with synthetic and organic nitrogen fertilizers thereby, causing a luxury consumption of nitrogen from the plants which results in the accumulation of nitrate (NO_3) in their vacuoles, and increasing the risks of nitrate contamination in groundwater and drinking water, harming both environment and human health. Nitrate itself is relatively non-toxic (Speijers 1996; Mensinga et al. 2003), while nitrite (NO_2), derived by nitrate bacterial reduction, and N-nitroso compounds, which are formed by reaction of nitrite with secondary nitrogen compounds such as amines (Fraser et al. 1980), before or after ingestion, are considered harmful for humans (Speijers 1996a; Speijers and van den Brandt 2003a). The potential harmful effects of nitrate-derived compounds on human health have prompted the development of government regulations and monitoring programs to protect public health by reducing the risk of nitrate accumulation in vegetables that along with drinking water and cured meat are considered the main source of nitrate intake (Walker 1990; Santamaria 1997).

This chapter focuses on the main features concerning the nitrate content in vegetables and the nitrate intake in the human diet, taking into account the recent EU Regulation No. 1258/2011 and the Scientific Opinion of the Panel on Contaminants in the Food chain (CONTAM) of the European Food Safety Authority (EFSA).

NITRATE CONTENT IN LEAFY VEGETABLES

Among the several factors affecting NO_3 accumulation in plants, the genetic component is one of the most important factors. Generally, nitrate accumulating vegetables belong to the families of *Brassicaceae* (rocket, radish, mustard), *Chenopodiaceae* (beetroot, Swiss chard, spinach), and *Amarantaceae* (amaranthus); but also *Asteraceae* (lettuce) and *Apiaceae* (celery, parsley) include species with high nitrate contents (Table 1.1). Nitrate content can also vary within species, cultivars, and even genotypes with different ploidy (Blom-Zandstra 1989).

Table 1.1: Classification of vegetables according to nitrate (NO_3) content (mg kg^{-1} fw)

Very low (< 200)	Low (200-500)	Middle (500-1,000)	High (1,000-2,500)	Very high (>2,500)
Artichoke	Broccoli	Cabbage	Celeriac	Celery
Asparagus	Carrot	'Cima di rapa'	Chinese	Chervil
Broad been	Cauliflower	(broccoli rab)	cabbage	Cress
Brussels sprouts	Cucumber	Dill	Endive	Lamb's lettuce
Eggplant	Pumpkin	'Radicchio'	Escarole	Lettuce
Garlic	'Puntarelle'	Savoy cabbage	Fennel	Radish
Onion	chicory	Turnip	Kohlrabi	Red beetroot
Green bean			Leaf chicory	Rocket
Melon			Leek	Spinach
Mushroom			Parsley	Swiss chard
Pea				
Pepper				
Potato				
Summer squash				
Sweet potato				
Tomato				
Watermelon				

The highest nitrate accumulating vegetable is rocket (Santamaria et al. 1999), a leafy vegetable popular in the Mediterranean region (Santamaria et al. 1998; 1998a; Santamaria et al. 2001). A number of species of the *Brassicaceae* family are grouped under the name of rocket belonging to the *Eruca* Miller and *Diplotaxis* DC genera. *Eruca* is present in both wild and cultivated forms; *Diplotaxis* is known as a wild type. In *Diplotaxis*, two surveys carried out in Italy showed that NO_3 content reaches up to 9,300 mg kg^{-1} (Cerutti et al. 1996; Santamaria et al. 1999). Rocket absorbs NO_3 very quickly (Santamaria et al. 1998; 1998a), and NO_3 concentration in leaves can be much higher than in the growing medium. In nutrient solution with 1 mM NO_3-N concentration, the NO_3 accumulation ratio (expressed as ratio between the concentration in leaves and in the nutrient solution) was, respectively, 55 for *E. vesicaria* and 101 for *D. tenuifolia* (Santamaria et al. 2002).

The different capacity to accumulate nitrate may be correlated with a different location of the nitrate-reductase activity (Pate 1973; Andrews 1986; Wallace 1986), as well as to a different degree of nitrate absorption and transfer in the plant (Maynard et al. 1976; Maynard and Barker 1979). Nitrate content differs in the various parts of a plant (Santamaria et al. 1999). Indeed, the vegetable organs can be listed by decreasing nitrate content as follows: petiole > leaf > stem > root > inflorescence > tuber > bulb > fruit > seed (Santamaria et al. 1999; Meah et al. 1994). Nitrate is mainly associated to cell vacuoles (Crawford and Glass 1998) and xylem transport system (Pate 1973), therefore, leafy vegetables accumulate higher levels of nitrate. Nitrate concentration in vegetables is also affected by storage conditions. In the event of consuming ready-to-use leafy vegetables, it is noteworthy that in vegetables stored at ambient temperature (22 °C) nitrate content may decrease, whereas nitrite tends to increase over time. Under refrigerated storage (7 days) at 5 °C, nitrate levels were almost unaffected in Chinese cabbage and spinach. Nitrite concentrations remained low over the whole storage period (Chung et al. 2004). This implies inactivation of endogenous nitrate reductase under cold storage conditions as well as prevention of bacterial activity, but only provided that cold chain is maintained.

Several surveys on nitrate concentration in vegetables have been conducted in Europe on behalf of the European Commission. A great amount of data on nitrate concentrations in vegetable samples provided to the European Commission by 20 EU Member States and Norway and covering the years from 2000 to 2007, were analyzed by EFSA. It is interesting to note that Italy, Austria, Portugal, Luxembourg, Malta, Poland, Bulgaria did not answer to this call. Nitrate median levels ranged from 1 mg kg^{-1} in some vegetables such as peas, up to 4,800 mg kg^{-1} for rocket, with the highest value reported for an oak leaf lettuce sample (19,925 mg kg^{-1}; Bottex et al. 2008). All the data were classified according to the main vegetable categories usually adopted in the horticultural production statistics. 'Herbs', 'leafy vegetables' and 'stem vegetables' showed the highest nitrate levels (Bottex et al. 2008). The group of the 'leafy vegetables' includes the most common species such as lettuce, spinach, Belgian endive, escarole and rocket, together with several minor species like dandelion and amaranth. In the group of 'herbs' were considered species, such as coriander, basil, borage, dill, which were showing high nitrate content in the median and in the 95th percentile values, however this is not alarming considering the small volume consumed of these species (EFSA 2008). Vice versa, although the total median concentration of nitrate for the 'roots and tubers' group is low (152 mg kg^{-1} fresh weight, fw), a particular concern is put in relation to potato and carrots which are both major components in the diet of many countries (EFSA 2008).

ACCEPTABLE DAILY INTAKE AND NITRATE INTAKE

The concept of the acceptable daily intake (ADI) is defined by the Joint Expert Committee of the Food and Agriculture (JECFA) Organization of the United Nations/World Health Organization (WHO) for substances intentionally added to food or for contaminants (pesticides, herbicides, and fertilizers; Speijers 1996; Speijers and van den Brandt 2003). In view of the well-known benefits of vegetables and the lack of data on the possible effects of vegetable matrices on the bio-availability of nitrate, JECFA considered it to be inappropriate to compare exposure to nitrate from vegetables with ADI or to derive limits for nitrate in vegetables directly from it (Speijers 1996). In the absence of an appropriate alternative approach in the literature, the consequences of nitrate intake exceeding the ADI are discussed.

The JECFA and the European Commission's Scientific Committee on Food (SCF) have set an ADI for NO_3 of 0-3.7 mg kg^{-1} body-weight (bw; SCF 1992; 1995; 1997;

Speijers 1996; Speijers and van den Brandt 2003). For a 60 kg adult, the ADI would correspond to a total intake of 222 mg of nitrate per day. The adopted values of ADI have been derived from long-term studies in rats and the sub-chronic toxicity study in dogs, given a no observable effect level (NOEL) of 500 mg kg^{-1} bw sodium nitrate, corresponding to 370 mg kg^{-1} bw nitrate and applying a safety factor of 100 (Speijers and van den Brandt 2003). The USA Environmental Protection Agency (EPA) Reference Dose (RfD) for nitrate is 1.6 mg nitrate nitrogen kg^{-1} bw day^{-1} (equivalent to about 7.0 mg NO_3 kg^{-1} bw day^{-1}) (Mensinga et al. 2003). The JECFA and SCF have proposed an ADI for NO_2 of 0-0.07 (Speijers 1996a; Speijers and van den Brandt 2003a) and 0-0.06 mg NO_2 kg^{-1} bw day^{-1} (SCF 1997), respectively, while the EPA has set an RfD of 0.1 mg nitrite nitrogen kg^{-1} bw day^{-1} (equivalent to 0.33 mg NO_2 kg^{-1} bw day^{-1}; Mensinga et al. 2003). The SCF retains that the ADIs are applicable to all sources of dietary exposure (SCF 1992; 1995).

Vegetables constitute the major dietary source of nitrate, generally providing from 30 to 94% of the daily dietary intake (Table 1.2). Their contribution to nitrite intake is low, and in fact lower than that from cured meat products (Walker 1990; Santamaria 1997; Lijinsky 1999). Nitrite is found in vegetables, typically in concentration of 1-2 mg kg^{-1} of fresh weight (Walker 1996). Potatoes, however, may contain up to 60 mg NO_2 kg^{-1} (Walker 1996). Higher amounts of nitrite are found in contaminated food or vegetables with broken tissues, stored for several days at room temperature (Lin and Yen 1980; Santamaria 1997). Compared with the current NO_3 and NO_2 SCFs ADIs, the ingestion of only 100 g of raw vegetables with a nitrate concentration of 2,500 mg kg^{-1} will already lead to an intake of 250 mg NO_3. Consuming this item alone, for a person of 60 kg, would exceed the ADI for nitrate by 13%. Considering the partial conversion of nitrate to nitrite (5%) after such consumption, the current SCF ADI for nitrite (0.06 mg kg^{-1} bw day^{-1}) would be exceeded by 247% (Santamaria 2006). A statistical exposure model has shown that in the adult population in the Netherlands 15% had nitrate daily intakes regularly exceeding the ADI; in young children this may rise to 45% (Slob et al. 1995).

In its last report, the SCF (1997) reviewed from a public health standpoint the presence of nitrate in foodstuffs, in general, and vegetables, in particular, and stated that the total intake of nitrate is normally well below the acceptable daily intake (Table 1.2). The major sources are potatoes and lettuce; the first because they are the vegetables consumed in largest quantity (Table 1.3), the latter due to its high nitrate content (SCF 1997). The SCF (1995, 1997) recommended continuation of efforts to reduce exposure to nitrate via food and water since nitrate can be converted into nitrite and nitrosamines, and urged that good agricultural practices should be adopted to ensure that nitrate levels are as low as reasonably achievable.

Table 1.2: Estimate of nitrate daily intake (drinking water not included), contribution of vegetables to nitrate daily intake and to ADI in various countries

Country	NO_3 (mg person^{-1})	Vegetable contribution[c] (%)	Reference
Belgium	148 [a]	93	Dejonckheere et al.(1994)
Denmark	61	64	Petersen and Stolze (1999)
European Union	18-131 [b]	100	SCF (1995)
Estonia	58 [b]	ND	Tamme et al. (2006)
Finland	77	92	Dich et al. (1996)
France	121	85	Cornée et al. (1992)
France	90	73	Menard et al. (2008)
Germany	68	72	Selenka and Brand-Grimm (1976)
Italy	149	90	CSS (1980)
Poland	65-85	ND	Borawska et al. (1998)
Spain	60	ND	ACAPV (1997)
Sweden	50	ND	Jagerstad et al. (1976)
The Netherlands	52	ND	Ellen et al. (1990)
United Kingdom	95	94	Knight et al. (1987)
United Kingdom	52	70	Ysart et al. (1999)
USA	73	90	NRC (1981)

[a] Included only vegetables and fruit;
[b] Included only vegetables;
[c] Not detected.

Table 1.3: Food consumption and nitrate daily intake in some countries of European Union (Schuddeboom, 1993)

Food group	Country	Food consumption (g person^{-1})	Average concentration (mg kg^{-1} fw)	NO_3 intake (mg person^{-1})	Contribution to NO_3 intake (%)
Potato	Germany	112	93	10.5	11.3
	Denmark	166	80	13.3	18.5
	The Netherlands	131	60	7.9	6.6
	United Kingdom	160	120	19.0	33.9
Other vegetables	Germany	73	721	52.6	56.4
	Denmark	114	440	50.2	69.7
	The Netherlands	150	800	120.0	85.7
	United Kingdom	162	136	22.0	39.3
Fruit	Germany	101	70	7.1	7.6
	Denmark	120	30	3.6	5.0
	The Netherlands	125	20	2.5	1.8
	United Kingdom	91	25	2.3	4.1

In previous research, nitrate intake from vegetables was estimated using the data on average daily per capita consumption of vegetables provided by the Italian

National Institute of Nutrition (Turrini et al. 1991) and on average nitrate content from the research work: nitrate daily intake from vegetables was 71 mg day^{-1}; over 30% of nitrate intake was derived from the consumption of lettuce and Swiss chard (Santamaria et al. 1999). In a more recent research, it was estimated that on average almost 66 mg of NO_3 day^{-1} (equivalent to 30% of the ADI) came from leafy vegetables, and green salad and lettuce were responsible for over 60% of calculated adult intake of nitrate from leafy vegetables (De Martin and Restani 2003). Chinese cabbage is considered the major daily nitrate and nitrite source of the Korean population. Taking into account its average nitrate content and per capita daily intake, it was estimated that an average nitrate intake from Chinese cabbage was 182 mg day^{-1} (80% of the NO_3 SCF ADI for a 60 kg bw person), and it was considered responsible for 60% of the nitrate intakes, being the most significant sources of nitrates (Chung et al. 2003).

Dietary exposures to nitrate for vegetarians are very similar to those of other consumers and are below the ADI. The average dietary exposure of the vegetarians in a UK study was 83 mg of NO_3 day^{-1} and the highest nitrate exposure was 209 mg day^{-1} (MAFF 1998, 1998a). Five exposure scenarios were hypothesized for nitrate intake from the EFSA Panel on Contaminants in Food Chain. In the first scenarios (S1), the consumption of 400 g of vegetables (levels found in dietary recommendations for the combined consumption of vegetables and fruit) could result in a nitrate exposure of 157 mg person^{-1} day^{-1}. High potato consumption (S2) could lead to nitrate exposure of 82 mg person^{-1} day^{-1} for a high percentile consumer. A high-level consumer of either spinach (S3A) or lettuce cultivars (S3B) could be exposed to nitrate levels of 104 and 178 mg person^{-1} day^{-1}, respectively. However, by replacing a third of the lettuce with rocket (S3C), the nitrate intake would rise up to 330 mg person^{-1} day^{-1}, doubling the contribution of nitrate compared to lettuce. In case a third of the vegetables used in S1 (400 g vegetable mix) would be consumed as spinach (S4A) or lettuce (S4B), the nitrate exposure would increase to 209 and 283 mg person^{-1} day^{-1}, respectively. Finally, a nitrate exposure of 337 and 457 mg person^{-1} day^{-1} for spinach (S5A) and lettuce cultivars (S5B), respectively, would be possible in regions characterized by the highest median nitrate levels. The various exposure scenarios demonstrate that the critical driver for a high dietary exposure to nitrate is not the absolute amount of vegetables consumed but the type of vegetable (i.e. leafy vegetables) and its nitrate concentration which is related to the production conditions. Leafy vegetables grown under less favourable (i.e. low solar radiation) conditions may potentially increase the nitrate dietary exposure by 50-60%. The contribution of potato consumption to nitrate dietary exposure is lower or much lower than the contribution from a mix of other vegetables (EFSA 2008) (Table 1.4).

Table 1.4: Various dietary exposure scenarios based on nitrate from vegetables only (EFSA, 2008)

Vegetable consumption [a] (g person^{-1} day^{-1})		Vegetable	Overall median nitrate concentration (mg kg^{-1})					Calculated exposure [b] (mg person^{-1} day^{-1})		
			Potato	Spinach	Lettuce	Rocket	Other	A	B	C
S1	400	Most	–	–	–	–	392	157	–	–
S2	771	Potato	106	–	–	–	–	82	–	–
S3	133 (44/89) [f]	Leafy	–	785	1,338	4,800	–	104 [d]	178 [e]	330 [f]
S4	133/267	Leafy/Most	–	785	1,338	–	392	209 [g]	283 [h]	–
S5	133/267	Leafy/Most	–	1,745 [c]	2,652 [c]	–	392	337 [g]	457 [h]	–

[a] S1: "recommended vegetable and fruit intake scenario" as 400 g person^{-1} day^{-1}; S2: "potato scenario"; S3: "green/leafy vegetables scenario"; S4: "combined S1 and S3 scenario"; S5: "regional scenario".

[b] A: spinach; B: all combined lettuce cultivars; C: a mix of rocket (1/3) and lettuce c (2/3).

[c] Highest regional median.

[d] Spinach at 133 g.

[e] Lettuce cultivars at 133.

[f] 1/3 of a leafy vegetable mix as rocket (44 g) and 2/3 as lettuce cultivars (89 g).

[g] Spinach at 133 g and other vegetables at 267 g.

[h] Lettuce cultivars at 133 g and other vegetables at 267 g.

It is worthy of note that most of the studies on daily nitrate exposure currently available are not comparable, first, because they are based on different methods of analysis, secondly because they take account of different daily consumption data (referring to general population or adults only), and finally, because some studies are focused on the whole diet, while some others take into consideration only part of the food sources of nitrate.

REGULATION AND GUIDELINES

To protect public health in response to the SCFs considerations of nitrate in food (SCF 1992, 1995, 1997), in 1997 the European Member States developed an EC Regulation setting limits for nitrate in lettuce and spinach (EC Regulation No. 194/97). The main purpose of this EC Regulation was to harmonise the limits of nitrate concentration in these vegetables in order to avoid trade difficulties across the European Union deriving from the different nitrate concentration limits in vegetables adopted in each Member State.

On 2 December 2011, the European Commission amended EC Regulation No. 194/97 (already amended with some periphrasis and integrations by EC Regulation No. 864/1999, 466/2001, 563/2002, 1822/2005 and 1881/2006) and adopted EC Regulation No. 1258/2011, setting new maximum levels for nitrate in some vegetables and baby foodstuffs. Actually the dispositions concerning nitrate content have been modified with respect to the previous EC Regulations as shown in Table 1.5. Since nitrate accumulation in vegetables is affected by environmental conditions, in particular temperature and solar radiation, the limits vary accordingly. Crops grown in the winter season have higher nitrate levels compared to those grown in the summer. Lower limits are fixed for open-grown lettuce than for lettuce grown in greenhouse, and in order to enable effective control the limits set for open-grown lettuce should apply also to lettuce grown under protected structures in the absence of precise labelling. The EC Regulation set also lower nitrate limits for "iceberg" lettuce compared to other lettuce types. The differences between nitrate levels in different cultivars have been most extensively studied in relation to lettuce where open leaf cultivars generally have higher nitrate concentrations than those characterized by tight headed such as iceberg (SCF, 1997).

Table 1.5: Maximum levels (limits) for nitrate (mg kg^{-1} fw) in some vegetables according to European Commission Regulation (EC) No. 1258/2011

Product	Harvest period	NO_3
Fresh spinach (*Spinacia oleracea* L)		3,500
Preserved, deep-frozen or frozen spinach		2,000
Fresh lettuce (*Lactuca sativa* L)	Harvested 1 October to 31 March:	
(protected and open-grown lettuce)	- lettuce grown under cover	5,000
excluding "Iceberg" type	- lettuce grown in the open air	4,000
	Harvested 1 April to 30 September:	
	- lettuce grown under cover	4,000
	- lettuce grown in the open air	3,000
"Iceberg" type lettuces	Lettuce grown under cover	2,500
	Lettuce grown in the open air	2,000
Rocket (*Eruca sativa*, *Diplotaxis* sp., *Brassica tenuifolia*, *Sisymbrium*)	Harvested 1 October to 31 March:	7,000
	Harvested 1 April to 30 September:	6,000
Baby foods and processed cereal-based foods for infants and young children		200

The EC Regulation takes into consideration that nitrate accumulation in vegetables is higher when solar radiation is lower. In fact, although similar nitrate concentration levels are reported in lettuces produced in Denmark, Belgium, Germany, Great Britain and the Netherlands, those values are significantly higher than those reported for lettuces grown in Italy and Spain, mainly due to the lower solar radiation conditions occurring in the former Member States (MAFF 1998b, 1998c). In reality, the different nitrate concentration limits adopted during the year derived from the political request advanced by the North European producers, which, considering the adverse environmental conditions, have a trade disadvantage compared to the South European colleagues.

In order to avoid any disadvantage, until 31 December, 2008, a derogation to the application of the EC Regulation No. 1881/2006 was in force in some Member States where the nitrate levels were reported to be frequently higher than those set in the EC Regulation. This derogation authorised these Member States "to place on the home market lettuce and/or spinach grown and intended for consumption in their territory". This derogation required annual monitoring to be carried out to demonstrate that nitrate levels in these crops were acceptable on public health grounds and that growers followed a "code of good agricultural practice". United Kingdom along with Ireland applied this derogation for both lettuce and spinach, while a derogation was allowed to Belgium and the Netherlands only for spinach. However, in these Member States, maximum limits do apply to lettuce and spinach imported from other Member States and third countries.

The recent EC Reg. No. 1258/2011 does not allow any more derogations for any of the Member States. However, to provide legal security for the Nord-European producers, that despite the adoption of good agricultural practices for reducing the nitrate content in spinach and lettuce as much as possible, often could not respect the nitrate levels set by the previous EU legislation for these crops, the new regulation has slightly increased the maximum level for nitrates in fresh spinach and lettuce. In more detail, the nitrate threshold has been raised to 3500 mg kg^{-1} fw regardless the harvesting period in fresh spinach, and in lettuce grown under cover and harvested in the winter-spring season (from 4500 up to 5000 mg kg^{-1} fw) or harvested in the spring-summer season (from 3500 up to 4000 mg kg^{-1} fw), as well as for lettuce grown in open field and harvested in the spring-summer season (from 2500 up to 3000 mg kg^{-1} fw) (Table 1.5).

Moreover, following the considerations expressed by the EFSA Panel for Contaminants in the Food Chain (EFSA 2008), the EC Reg. No. 1258/2011 has set new limits of the nitrate content also in different rocket species such as *Eruca sativa*, *Diplotaxis* spp., *Brassica tenuifolia* and *Sisymbrium tenuifolium*. Given the high nitrate levels often found in these crops, maximum thresholds of 7000 and 6000 mg kg^{-1} fw have been set for rocket harvested from October 1 to March 31 and from April 1 to September 30, respectively (Table 1.5). Such limits are far less stringent than those imposed for rocket in Switzerland and in Germany, before the adoption of the EC Reg. No. 1258/2011. Namely, before publication of the last EC Regulation, nitrate content for rocket destined to these markets, could not exceed thresholds of 2500-4000 mg kg^{-1} fw, which were considered restrictive and difficult to respect even, under favourable environmental (sunlight) conditions or when reduced nitrate fertilizer rates are used for this crop (Bianco et al. 1998).

Beside the EU Regulation, maximum limits of nitrate content have been imposed for other vegetables trade in some European countries. 'Guidelines' for nitrate content of potatoes have been developed in Germany and Poland where, only tubers with less than 200 and 183 mg kg^{-1} fw are accepted respectively (Cieslik and Sikora 1998).

National limits for several vegetables were also adopted in Estonia before joining the EU in 2004 (Tamme et al. 2006). Maximum levels of nitrate content are also imposed for the export in Russia of several vegetables such as lettuce (from

2500 to 4500 mg kg^{-1} fw), cabbage (from 500 to 900 mg kg^{-1} fw) and other leafy vegetables (2000 mg kg^{-1} fw), tomato (from 150 to 300 mg kg^{-1} fw), cucumbers (from 150 to 400 mg kg^{-1} fw) and other cucurbits, potatoes (250 mg kg^{-1} fw), carrots (from 250 to 400 mg kg^{-1} fw) and onion (80 mg kg^{-1} fw) (Anonymous 2010). Although, no nitrate vegetables content limits have been introduced in the USA and in Brazil (Guadagnin et al. 2005), the nitrate content in vegetables is subject to continuous monitoring. In the same way, it is not considered necessary to establish limits of nitrate vegetables contents in South Korea (Chung et al. 2003), and in general in the South-East Asian countries. A maximum vegetable nitrate level (3100 mg kg^{-1} fw) has been established instead in China (Zhou et al. 2000).

Considering the higher risks for nitrate intake in infants, on 7 April 2004, the EC Regulation No. 655/2004 established 200 mg kg^{-1} on 'as sold' basis, the maximum level of nitrate allowed in baby foods and processed cereal-based foods for infants and young children. The same limit has been confirmed by the recent EC Reg. No. 1258/2011. As regards the methods of analysis of the nitrate content, while an official method of analysis for nitrate has been adopted from EPA in USA, no official method has been published in the EU legislation and nitrate levels in vegetables are generally assayed by modifying the protocols used for other food typologies. The recent EC Regulation No. 1882/2006 laid down the methods of sampling and extraction, but gives only so-called 'criteria approach' for the choice of the analysis method, without prescribing a specific official method. The same Regulation also sets performance criteria for recovery and precision, that should be met by all analytical methods used. Recommended recovery values for concentrations < 500 mg kg^{-1} and $= 500$ mg kg^{-1} are set as 60-120% and 90-110%, respectively. A study funded from the UK Food Standard Agency on extraction procedures for measuring nitrate in lettuce and spinach indicates that a method based on BS EN 12014-2 should be used, involving hot water extraction, while the choice of chromatographic detection method (HPLC or IC) rather than colorimetry is recommended (Farrington et al. 2006).

CONCLUSION

In the holistic analysis of the problem related to the human nitrate intake through the vegetable daily consumption, the CONTAM panel adopts the risk/benefit assessment criteria. On one hand, the CONTAM takes into account the presence in vegetables of nitrate and other anti-nutrients such as allergens with potential harmful effect on human health. On the other hand, it highlights the potential defence role of nitrate, as nitric oxide, against enteropathogens and inflammations (Lundberg et al. 2006, 2008), and the presence in vegetables of bioactive substances and a high content of micro and macronutrients that may have beneficial health effects (Bottex et al. 2008). The CONTAM concludes that, in spite of the daily 400 g fruits and/or vegetables ration recommended from WHO, only a minority of the European population has a daily consumption of 400 g of vegetables, and even in this case the nitrate ADI is seldom exceeded. In our opinion, even considering the low percentage of consumers adopting a diet characterized by high leafy vegetable intake, the reduction of dietary nitrate intake should be recommended as a preventive health care measure. Vegetable nitrate content should be kept, therefore, as low as possible, by ensuring the adoption (volunteer or mandatory) from the growers, of good agricultural practices. A reduction in nitrate content may represent an added value for vegetable products already rich in bioactive compounds such as carotenoids, vitamins C and E, selenium, dietary fibre, plant sterols, glucosinolates, indoles, isothiocyanates, flavonoids, phenols, etc.

REFERENCES

ACAPV (Administración de la Communidad Autónoma del País Vasco), (1997). Food Chemical Surveillance in the Basque Country: 1990-1995 (Vitoria-Gasteiz: Servicio Central de publicaciones del Gobierno Vasco).

Andrews, M. (1986). The partitioning of nitrate assimilation between the root and shoot of higher plants. Mini-review. *Plant Cell Environ.* **9:** 511-519.

Anonymous. (2010). MRLs Nitrate Russia. Available from http://ewrs.rivm.nl/object_binary/o4555_RussischeMRLsNitratev4.pdf (accessed May 16, 2013).

Bianco, V.V., Santamaria, P. and Elia, A. (1998). Nutritional value and nitrate content in edible wild species used in southern Italy. *Acta Hortic.* **467:** 71-87.

Blom-Zandstra, M. (1989). Nitrate accumulation in vegetables and its relationship to quality. *Ann. Appl. Biol.* **115:** 553-556.

Borawska, M., Markiewica, R. and Wikowska, A. (1998). Nitrate and nitrite content in daily hospital diets during the winter season—comparison of analytical and calculation methods. *Eur. J. Clinical Nutr.* **41:** 489-493.

Bottex, B., Dorne, J.L.C.M., Carlander, D., Benford, D., Przyrembel, H., Heppner, C., Kleiner, J. and Cockburn, A. (2008). Risk-benefit health assessment of food—Food fortification and nitrate in vegetables. *Trends Food Sci. Tech.* **19:** S113-S119.

Cerutti, G., Bertolé, S., Casartelli, A. and D'Amato, A. (1996). Nitrati e nitriti in ortaggi e aromatizzanti vegetali del commercio. *La difesa delle piante* **19:** 35-40.

Chung, J-C., Chou, S-S. and Hwang, D-F. (2004). Changes in nitrate and nitrite of four vegetables during storage at refrigerated and ambient temperatures. *Food Addit. Contam.* **21:** 317-322.

Chung, S.Y., Kim, J.S., Kim, M., Hong, M.K., Lee, J.O., Kim, C.M. and Song, I.S. (2003). Survey of nitrate and nitrite contents of vegetables grown in Korea. *Food Addit. Contam.* **20:** 621-628.

Cieslik, E. and Sikora, E. (1998). Correlation between the levels of nitrates and nitrites and the contents of potassium, calcium and magnesium in potato tubers. *Food Chem.* **63:** 25-528.

Cornée, J., Lairon, D., Velema, J., Guyader, M. and Berthezene, P. (1992). An estimate of nitrate, nitrite and N-nitrosodimethylamine concentrations in French food products or food groups. *Sci. Alim.* **12:** 155-197.

Crawford, N.M. and Glass, A.D.M. (1998). Molecular and physiological aspects of nitrate uptake in plants. *Trends Plant Sci.* **3:** 389-395.

CSS (Consiglio Superiore Sanità), (1980). Commissione "Nitriti, nitrati e nitrosamine". Documento conclusivo, Rome, p. 154.

De Martin, S. and Restani, P. (2003). Determination of nitrates by a novel ion chromatographic method: occurrence in leafy vegetables (organic and conventional) and exposure assessement for Italian consumers. *Food Addit. Contam.* **20:** 787-792.

Dejonckheere, W., Steurbaut, W., Drieghe, S., Verstraeten, R. and Braeckman, H. (1994). Nitrate in food commodities of vegetable origin and the total diet in Belgium (1992-1993). *Microbiol. Alim. Nutr.* **12:** 359-370.

Dich, J., Järvinen, R., Knekt, P. and Penttilä, P.L. (1996). Dietary intakes of nitrate, nitrite and NDMA in the Finnish mobile clinic health examination survey. *Food Addit. Contam.* **13:** 541-552.

EC (European Commission) (2011) Commission Regulation (EC) No. 1258/2011 of 2 December 2011 amending Regulation (EC) No 1881/2006 as regards maximum levels for nitrates in foodstuffs. *Official J European Communities* L320:15-17.

EC (European Commission), (1997). Commission Regulation (EC) No. 194/97 of 31 January 1997 setting maximum levels for certain contaminants in foodstuffs. *Official J. Eur. Commun.* **L31:** 48-50.

EC (European Commission), (1999). Commission Regulation (EC) No. 864/99 of 26 April 1999 amending Regulation (EC) No. 194/97 setting maximum levels for certain contaminants in foodstuffs. *Official J. Eur. Commun.* **L108:** 16-18.

EC (European Commission), (2001). Commission Regulation (EC) No. 466/2001 of 8 March 2001 setting maximum levels for certain contaminants in foodstuffs. *Official J. Eur. Commun.* **L77:** 1-13.

EC (European Commission), (2002). Commission Regulation (EC) No. 563/2002 of 2 April 2002 amending Regulation (EC) No. 466/2001 setting maximum levels for certain contaminants in foodstuffs. *Official J. Eur. Commun.* **L86:** 5-6.

EC (European Commission), (2004). Commission Regulation (EC) No. 655/2004 of 7 April 2004 amending Regulation (EC) No. 466/2001 as regards nitrate in foods for infants and young children. *Official J. Eur. Commun.* **L104:** 48-49.

EC (European Commission), (2006). Commission Regulation (EC) No. 1881/2006 of 19 December 2006 setting maximum levels for certain contaminants in foodstuffs. *Official J. Eur. Commun.* **L364:** 5-24.

EC (European Commission), (2006). Commission Regulation (EC) No. 1882/2006 of 19 December 2006 laying down methods of sampling and analysis for the official control of level of nitrate in certain foodstuffs. *Official J. Eur. Commun.* **L364:** 25-31.

EFSA (European Food Safety Authority), (2008). Nitrate in vegetables. Scientific Opinion of the Panel on Contaminants in the Food Chain. EFSA Journal 689:1-79. Available from http://www.efsa.europa.eu/EFSA/Scientific_Opinion/contam_ej_689_nitrate_en.pdf (accessed May 16, 2013).

Ellen, G., Egmond, E., van Loon, J.W., Sahertian, E.T. and Tolsma, K. (1990). Dietary intakes of some essential trace elements, nitrate, nitrite and N-nitrosamines, by Dutch adults: estimated via a 24-hour duplicate portion study. *Food Addit. Contam.* **7:** 207-221.

Farrington, D., Damant, A.P., Powell, K., Ridsdale, J., Walker, M. and Wood, R. (2006). A comparison of the extraction methods used in the UK nitrate residues monitoring program. *J. Assoc. Public Anal.* **34:** 1-11.

Fraser, P., Chilvers, C., Beral, V. and Hill, M.J. (1980). Nitrate and human cancer: a review of the evidence. *Int. J. Epidemiol.* **9:** 3-11.

Gangolli, S.D., van den Brandt, P., Feron, V., Janzowsky, C., Koeman, J., Speijers, G., Speigelhalder, B., Walker, R. and Winshnok, J. (1994). Assessment of nitrate, nitrite, and N-nitroso compounds. *Eur. J. Pharmacol. Environ. Toxicol. Pharmacol. Sect.* **292:** 1-38.

Guadagnin, S.G., Rath, S. and Reyes, F.G.R. (2005). Evaluation of nitrate content in leafy vegetables produced through different agricultural systems. *Food Addit. Contam.* **22:** 1203-1208.

Hambridge, T. (2003). Nitrate and nitrite. Intake assessment. WHO Food Additives Series 50, http://www.inchem.org/documents/jecfa/jecmono/v50je07.htm (accessed May 16, 2013).

Jagerstad, M. and Nilsson, R. (1976). Intake of nitrate and nitrite of some Swedish consumers as measured by the duplicate portion technique, in Proc. Sec. Symp. Nitrite Meat Prod., Ed by Jagerstad, M. and Nilsson, R. Zeist, pp. 283-287.

Knight, T.M., Forman, D., Al-Dabbagh, S.A. and Doll, R. (1987). Estimation of dietary intake of nitrate and nitrite in Great Britain. *Food Chem. Toxic.* **25:** 277-285.

Lijinsky, W. (1999). N-Nitroso compounds in the diet. Mut. Res. 443: 129-138.

Lin, J-K. and Yen, J-Y. (1980). Changes in the nitrate and nitrite contents of fresh vegetables during cultivation and post-harvest storage. *Fd. Cosmet. Toxicol.* **18:** 597-603.

Lundberg, J.O., Feelisch, M., Björne, H., Jansson, E.Å. and Weitzberg, E. (2006). Cardioprotective effects of vegetables: Is nitrate the answer? *Nitric Oxide* **15:** 359-362.

Lundberg, J.O., Weitzberg, E. and Gladwin, M.T. (2008). The nitrate-nitrite-nitric oxide pathway in physiology and therapeutics. *Nat. Rev. Drug. Discov.* **7:** 156-167.

MAFF (Ministry of Agriculture, Fisheries and Food), (1998). 1997 total diet study—nitrate and nitrite. Food surveillance information sheet N. 163, MAFF, London, http://archive.food.gov.uk/maff/archive/food/infsheet/1998/no163/163tds.htm (accessed May 16, 2013).

MAFF (Ministry of Agriculture, Fisheries and Food), (1998a). Duplicate diet study of vegetarians—Nitrate analysis. Food surveillance information sheet N. 165, MAFF, London, http://archive.food.gov.uk/maff/archive/food/infsheet/1998/no165/165dds.htm (accessed May 16, 2013).

MAFF (Ministry of Agriculture, Fisheries and Food), (1998b). 1997/98 UK monitoring programme for nitrate in lettuce and spinach. Food surveillance information sheet N. 154, MAFF, London, http://archive.food.gov.uk/maff/archive/food/infsheet/1998/no154/154lettu.htm (accessed May 16, 2013).

MAFF (Ministry of Agriculture, Fisheries and Food), (1998c). UK monitoring programme for nitrate in lettuce and spinach. Food surveillance information sheet N. 16/01, MAFF, London, http://archive.food.gov.uk/fsainfsheet/2001/no16/16nitrate.htm (accessed May 16, 2013).

Maynard, D.N. and Barker, A.V. (1979). Regulation of nitrate accumulation in vegetables. *Acta Hortic.* **93:** 153-162.

Maynard, D.N., Baker, A.V., Minotti, P.L. and Peck, N.H. (1976). Nitrate accumulation in vegetables. *Adv. Agron.* **28:** 71-118.

Meah, M.N., Harrison, N. and Davis, A. (1994). Nitrate and nitrite in foods and the diet. *Food Addit. Contam.* **11:** 519-532.

Menard, C., Heraud, F., Volatier, J-L. and Leblanc, J-C. (2008). Assessment of dietary exposure of nitrate and nitrite in France. *Food Addit. Contam. Part A* **25:** 8971-8988.

Mensinga, T.T., Speijers, G.J.A. and Meulenbelt. J. (2003). Health implications of exposure to environmental nitrogenous compounds. *Toxicol. Rev.* **22:** 41-51.

NRC (National Research Council Committee on Nitrite and Alternative Curing Agents in Food), (1981). The health effects of nitrates, nitrite, and N-nitroso compounds. National Academy Press, Washington.

Pate, J.S. (1973). Uptake, assimilation and transport of nitrogen compounds by plants. *Soil Biol. Biochem.* **5:** 109-119.

Petersen, A. and Stoltze, S. (1999). Nitrate and nitrite in vegetables on the Danish market: content and intake. *Food Add. Contam.* **16:** 291-299.

Santamaria, P. (1997). Contributo degli ortaggi all'assunzione giornaliera di nitrato, nitrito e nitrosammine. *Industrie Alimentari* **36:** 1329-1334.

Santamaria, P. (2006). Nitrate in vegetables: toxicity, content, intake and EC regulation. *J. Sci. Food Agric.* **86:** 10-17.

Santamaria, P., Elia, A. and Serio, F. (2002). Effect of solution nitrogen concentration on yield, leaf element content, and water and nitrogen use efficiency of three hydroponically-grown rocket salad genotypes. *J. Plant Nutr.* **25:** 245-258.

Santamaria, P., Elia, A., Gonnella, M., Parente, A. and Serio, F. (2001). Ways of reducing rocket salad nitrate content. *Acta Hortic.* **548:** 529-537.

Santamaria, P., Elia, A., Papa, G. and Serio, F. (1998). Nitrate and ammonium nutrition in chicory and rocket salad plants. *J. Plant Nutr.* **21:** 1779-1789.

Santamaria, P., Elia, A., Parente, A. and Serio, F. (1998a). Fertilization strategies for lowering nitrate content in leafy vegetables. Chicory and rocket salad cases. *J. Plant Nutr.* **21:** 1791-1803.

Santamaria, P., Elia, A., Serio, F. and Todaro, E. (1999). A survey of nitrate and oxalate content in retail fresh vegetables. *J. Sci. Food Agric.* **79:** 1882-1888.

SCF (Scientific Committee on Food), (1992). Opinion on nitrate and nitrite, expressed on 19 October 1990 (26th series), Ed by European Commission. Brussels, pp. 23-32.

SCF (Scientific Committee on Food), (1995). Opinion on nitrate and nitrite, expressed on 22 September 1995 (Annex 4 to Document III/5611/95), Ed by European Commission. Brussels, pp. 20.

SCF (Scientific Committee on Food), (1997). Assessment of dietary intake of nitrates by the population in the European Union, as a consequence of the consumption of vegetables, in Reports on tasks for scientific cooperation. Report of experts partecipating in Task 3.2.3, Ed by European Commission. Brussels, pp. 34.

Schuddeboom, L.J. (1993). Nitrates and nitrites in foodstuffs, Ed by European Commission. Council of Europe Press, Brussels, p. 125.

Selenka, F. and Bran-Grimm, D. (1976). Nitrat und nitrit in der Ernährung des Meuschen. Kalkulation der mittleren Tagesonfuahme und Abschätzung der Schawankungsbreite. *Zbl Bakt Hyg I Abt Orig B* **162:** 449-466.

Slob, W., van der Berg, R. and van Veen, M.P. (1995). A statistical exposure model applied to nitrate intake in the Dutch population, in Health aspects of nitrates and its metabolites (particularly nitrite), Ed by European Commission. Council of Europe Press, Strasbourg, pp. 75-82.

Speijers, G.J.A. (1996). Nitrate, in Toxicological evaluation of certain food additives and contaminants in food, Ed by WHO (World Health Organization). Food Additives Series 35, Geneva, pp. 325-360.

Speijers, G.J.A. (1996a). Nitrite (and potential endogenous formation of N-nitroso compounds), in Toxicological evaluation of certain food additives and contaminants in food, Ed by WHO (World Health Organization). Food Additives Series 35, Geneva, pp. 269-323.

Speijers, G.J.A. and van den Brandt, P.A. (2003). Nitrate (and potential endogenous formation of N-nitroso compounds), WHO Food Additives Series 50, http://www.inchem.org/documents/jecfa/jecmono/v50je06.htm (accessed May 16, 2013).

Speijers, G.J.A. and van den Brandt, P.A. (2003) Nitrite (and potential endogenous formation of N-nitroso compounds), WHO Food Additives Series 50, http://www.inchem.org/documents/jecfa/jecmono/v50je05.htm (accessed May 16, 2013).

Tamme, T., Reinik, M., Roasto, M., Juhkam, K., Tenno, T. and Kiis, A. (2006). Nitrates and nitrites in vegetables and vegetable-based products and their intakes by the Estonian population. Food Addit. Contam. **23:** 355-361.

Turrini, A., Saba, A. and Lintas, C. (1991). Study of the italian reference diet for monitoring food constituents and contaminants. *Nutr. Res.* **11:** 861-873.

Walker, R. (1990). Nitrates, nitrites and N-nitroso compounds: a review of the occurrence in food and diet and the toxicological implications. *Food Addit. Contam.* **7:** 717-768.

Walker, R. (1996). The metabolism of dietary nitrites and nitrates. *Biochem. Soc. Trans.* **24:** 780-785.

Wallace, W. (1986). Distribution of nitrate assimilation between the root and shoot of legumes and a comparison with wheat. *Physiol. Plant* **66:** 630-636.

Ysart, G., Miller, P., Barrett, G., Farrington, D., Laurance, P. and Harrison, N. (1999). Dietary exposure to nitrate in the UK. *Food Addit. Contam.* **16:** 521-532.

Zhou, Z.Y., Wang, M.J. and Wang, J.S. (2000). Nitrate and nitrite contamination in vegetables in China. *Food Rev. Int.* **16:** 61-76.

2

Nitrate and Edible Plants in the Mediterranean Region of Turkey: An Overview

Münir Öztürk[1], **Salih Gücel**[2], **Serdal Sakçali**[3] **and Süleyman Baslar**[4]

[1]Ege University, Botany Department, Bornova, Izmir, Turkey
[2]Near East University, Environmental Sciences Institute, Nicosia, Cyprus
[3]Fatih University, Biology Department, Istanbul,Turkey
[4]Faculty of Education, Dokuzeylül University, Buca- Izmir, Turkey
e-mail: munirozturk@gmail.com

ABSTRACT

Turkey is divided in three main phytogeographical regions; one of these is the Mediterranean. The total gene pool of wild species and land races of cultivated plants has a great potential in the plant cover of this region. The studies undertaken in this region revealed that large levels of fertilizers, in particular nitrates, are used in thousands of greenhouses and on the farms, most of these being used unknowingly. A large number of producers apply both inorganic as well as organic fertilizers. The amount varies between 40–75 kg/da fertilizer, which is tenfold of the average recommended for the area. The excessive use of these fertilizers under long periods of drought to leads accumulation of nitrates in plants, which in turn poses a great threat as a potential pollution source for the soils and waters of the region. This paper presents an overview of the results published on the nitrate pollution and accumulation in the plants of this region.

Keywords: Turkey, Mediterranean region, Nitrates, Fertilizers

INTRODUCTION

Nitrogen as an important part of plants is one of the most abundant elements on the earth, but its deficiency is probably the most common nutritional problem affecting plant growth globally. The nitrogen in soil that might eventually be used by plants has two sources—nitrogen containing minerals and the vast storehouse of nitrogen in the atmosphere. The nitrogen in soil minerals is released as the mineral decomposes. This process is generally quite slow and contributes only slightly to nitrogen nutrition on most soils. In the soil it exists in three general forms—organic nitrogen compounds, ammonium (NH_4^+) ions, and nitrate (NO_3^-) ions. At any given time, nearly 96 percent of the potentially available nitrogen in the soil is in organic form either in plants or animal residues (manure, residues of forage legumes, etc.), relatively stable soil organic matter, or living soil organisms, mainly microbes such as bacteria. This nitrogen is not directly available to plants, but some can be converted to available forms by microorganisms (Fig. 2.1) (Ozturk and Secmen 1995). The concentration of this essential element in the plants can be as high as 4–5 percent of the dry matter. It is impossible to generalise the profitability of use of nitrogen fertilizers. These are used to help fill the feed gaps on large number of farmlands. However, it is necessarry to establish if there is a need for extra pasture to feed the animals before using nitrogen. Application of nitrogen fertilizers to the pastures is an important management practice at many places. The principles governing their use are not affected by differences between the pastures, but differences in growth patterns can prove effective to increase the production.

Fertile soils contain large amounts of nitrogen in plant and animal residues, animal excreta and humified soil (Whitehead 1995). This nitrogen is not taken up by the plants. But microbial breakdown of these leads to the plant available nitrogen with its release in ammonium forms through mineralization, and a part is converted to nitrate forms (nitrification). These processes are influenced by both soil water availability and temperature. The amount of plant available N in soil at any one time (the ammonium and nitrate pools) is usually relatively low under irrigated conditions, at 10–15 kg N/ha in the 0–30 cm horizon. The majority of this is in the ammonium form and the amount of nitrate can be as low as 1 kg N/ha. The rate of mineralisation of organic-N in soils can limit the plant growth. When the available N supply is less than that required for potential grass growth, the plant will respond to N fertilizer.

The quantity of nitrogen released from any of these materials depends on the composition of the material, particularly its ratio of carbon to nitrogen, and the weather. When microorganisms decompose manure and alfalfa residue having C/N ratios of 10:1 to 20:1, net mineralization occurs because the microbes are releasing nitrogen from the residue faster than they can utilize it (Guleryuz et al. 2009). On soils containing large quantities of NH_4-rich clays, however, nitrogen supplied by the mineral fraction may be significant in some years.

A very small amount of organic nitrogen may exist in soluble organic compounds such as urea that may be slightly available to plants. The majority of plant-available nitrogen is in the inorganic NH_4^+ and NO_3-forms. Ammonium ions bind to the negatively charged cation exchange complex of the soil and behave much like other cations in the soil. Nitrate ions do not bind to the soil solids because they carry negative charges, but exist dissolved in the soil water, or precipitated as soluble salts under dry conditions. Some NH_4^+ and NO_3^- may also exist in the crystal structure of certain soil minerals, and may be quite available.

Immobilization and mineralization by microorganisms are most rapid when soils are warm and moist, but not saturated with water. The quantity of inorganic nitrogen available for crop use often depends on the amount of mineralization occurring, and the balance between mineralization and immobilization (Guleryuz et al. 2009). The nitrogen cycle contains several routes by which plant-available nitrogen can be lost from the soil. Nitrate-nitrogen is usually more subject to loss than ammonium-nitrogen. Significant loss mechanisms include leaching, denitrification, volatilization, and crop Fig. 2.1. The nitrate form of nitrogen is so soluble that it leaches easily when excess water percolates through the soil. This can be a major loss mechanism in coarse-textured soils where water percolates freely, but is less of a problem in finer-textured, more impermeable soils, where percolation is very slow. Denitrification can cause major losses of nitrogen when soils are warm and remain saturated for more than a few days. Crop removal represents a loss because nitrogen in the harvested portions of the crop plant is removed from the field completely. The nitrogen in crop residues is recycled back into the system and is better thought of as immobilized rather than removed.

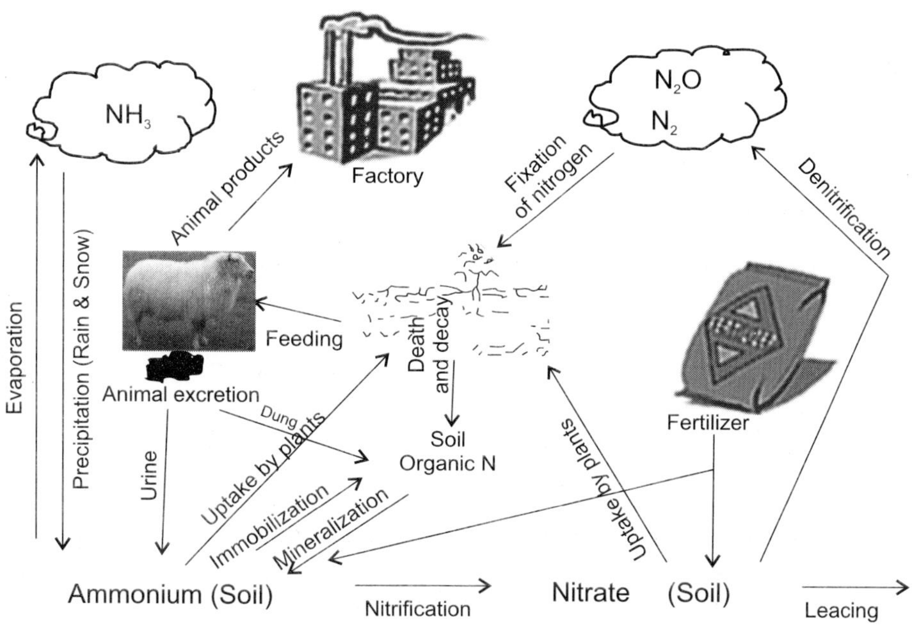

Fig. 2.1. A schematic representation of the nitrogen cycle (modified from Öztürk ve Seçmen, 1995).

Anthropogenic activities aimed at enhancing food production facilitate accumulation of nitrates in plants and affect the quality of soil and water resources adversely. The reason being when input of nitrogen exceeds the demand, plants are no longer able to absorb it, and nitrogen builds up in the soil, mostly as nitrates. This causes imbalance of nutrients in the soil and increases the nitrate level in groundwater supplies, which influences the nitrate content of plants, especially the leafy vegetables (Vieira et al. 1998; Dapoigny et al. 2000; Nosengo 2003; NAAS 2005; Anjana et al. 2007 a,b,c). At the same time, long periods of drought also result in an accumulation of nitrates in plants. When more soil nitrogen is present than that needed for maximum growth, some plants tend to accumulate nitrate even without environmental stress. This response is particularly true with hardy soil feeders like sorghum, noted for "luxury consumption" of certain nutrients.

Farmers generally apply excessive amounts of nitrogenous fertilizers to their crops as an insurance against yield losses. The situation in Turkey is not different in particular in the mediterranean phytogeographical region, which is characterized by dry hot summers and moderate winters, with low precipitation. The soils are a mixture of old and newer soils, but distinctive terra rossa soil is widely distributed in the region. The bare and steep slopes, broad floodplains and torrential forms are the main kinds of landforms. It is an area of megadiversity of important food crops, pasture and range land species. The total gene pool of wild species and land races of cultivated plants has a great potential in its flora. Excessive levels of nitrogen in the form of nitrate have been found in some groundwaters in the region. Well waters, especially from shallow wells, are sometimes high in nitrates. Similarly, runoff from feedlot grounds is reported to be high in nitrite (Atilgan et al. 2007). The aim of this review is thus to present an overview of the current situation on the nitrate accumulation in the plants in the Mediterranean part of Turkey.

FERTILIZERS AND NITRATES

The role of fertilizers came to forefront in the 19th century. After the discovery of potassium and phosphorus as the key elements in the growth of plants, these were easily added to the soil due to their presence in huge amounts on earth. However, the same was not true for nitrogen. The programme for the synthetic production of nitrogen began, and for the first time, Fritz Haber, a Chemist produced ammonia from the atmospheric nitrogen and hydrogen. He was awarded a Nobel Prize in 1918. The industrial production of nitrogen fertilizer was given a start through the joint efforts of Karl Bosch and Haber, and in 1913, a factory with the production capacity of 60,000 tons was established in Germany. This method is still used in the production of nitrogenous fertilizers. Today, the amount of nitrogenous fertilizers produced has reached a level of around 175 million tons. The developments in this direction followed the trend in Turkey. The first fertilizer production was started in 1939 at Karabuk. In this iron and steel factory, sulfuric acid and ammonia gas were used to produce ammonium sulfate (21%N), followed by the production of super phosphate. An impetus was given to the production from 1954 onwards when nitrogen industry was developed with facilities in Kutahya, Samsun, Gemlik, and Elazig. In 1984, the name of this industry was changed to Turkey Fertilizer Industry Inc. (TÜGSAŞ) as an entire state-owned institution. However, with the decision of the Supreme Council of Privatization, TÜGSAŞ was privatized in 1998, but in the year 2000, Elazig Fertilizer Industries Inc. was transferred to Firat University. Presently, the total production is nearly six million tons. Out of this 35–40 percent is provided by TÜGSAŞ, 25 percent by Toros Fertilizers, and 15 percent by Gübretaş. The import of fertilizers increased after the year 2000 at a level of 110 percent as compared to 1996, but has decreased to 3 percent after 2005. Toros Tarim Mersin Factory (Figs. 2.2, 2.3) is a fertilizer production complex producing ammonium nitrate (26–33 percent nitrogen) and diamonium phosphate (18 percent nitrogen and 46 percent P_2O_5). In the ammonium nitrate unit, calcium ammonium nitrate (CAN) fertilizer with 26 percent nitrogen is produced together with ammonium nitrate with 33 percent nitrogen. In the main production units, 5,94,000 tons/year CAN and 1,50,000 tons/year DAP are produced. It produces 38 percent of CAN, and 28 percent of DAP out of the countries' total production.

Fig. 2.2. A general view of Toros Tarim Mersin Factory alongside the Mediterranean coast in Turkey.

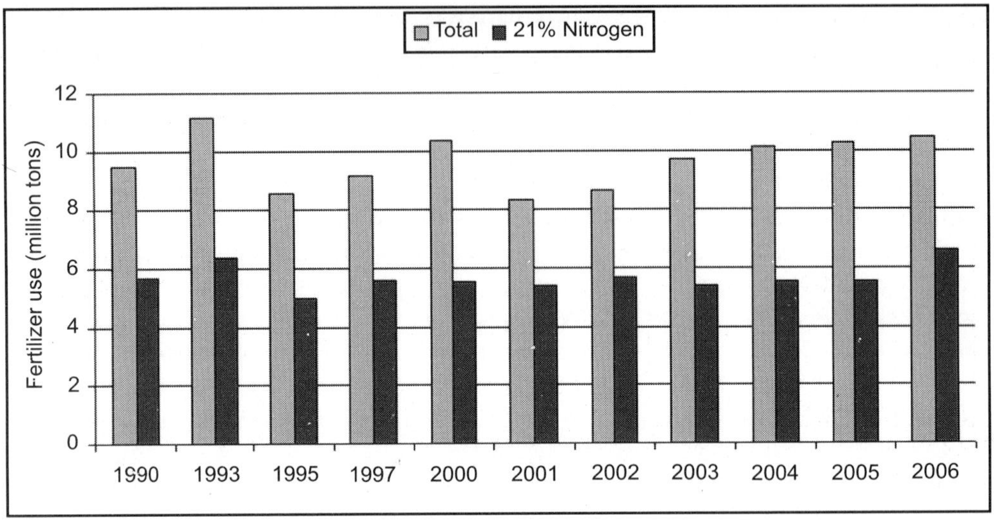

Fig. 2.3. Fertilizer use (tons) (Ministry of Agriculture and Rural Affairs—On the basis of Sales).

Fertilizers like urea, ammonium nitrate and diammonium phosphate (DAP) have single chemical compound with more than one element, and mixed ones like nitrogen, phosphorus and potassium blends, nitrogen and phosphorus blends and nitrogen and nitrogen blends have more than one chemical compound. They have different contents and ratios of nitrogen (Fig. 2.4).

Nitrogenous fertilizers are the most important input in agicultural production. Several reports suggest that some forms of N are more effective for plant growth than others. Annually, the N_2O levels in the atmosphere increase by 0.2–0.3 percent globally (Bacon 1995; Abdalla et al. 1998). Asian countries consume 48 percent of the total fertilizer production in the world (Yilmaz 2004). When applied insufficiently, losses in yield and quality occur, however, overusage causes washing of nitrogen and phosphorous fertilizers away, leading to the pollution of surface and groundwaters and air pollution with the emission of nitrogen oxides (Guler 2004; Atilgan et al. 2007). All nitrogen materials are usually manufactured from ammonia. Such materials are less expensive, more concentrated, and just as plant-available as the organics used in the past. The readily available soluble forms of nitrogen are anhydrous ammonia and aqua ammonia, ammonium nitrate, urea, non-pressure nitrogen solutions, ammonium sulfate, and ammonium phosphates. The rate at which these should be applied depends on the weather and soil conditions. When the soils are warm and moist, decomposition proceeds rapidly, and nitrogen released may be significant, but when the soils are cold and wet, or very dry, nitrogen release may be very much less than expected.

There are two major types of fertilizers consumed in the Mediterranean region of Turkey. These are solid soil chemical fertilizers and liquid leaf fertilizers. The former includes ammonium sulphate, ammonium nitrate, diammonium phosphate,

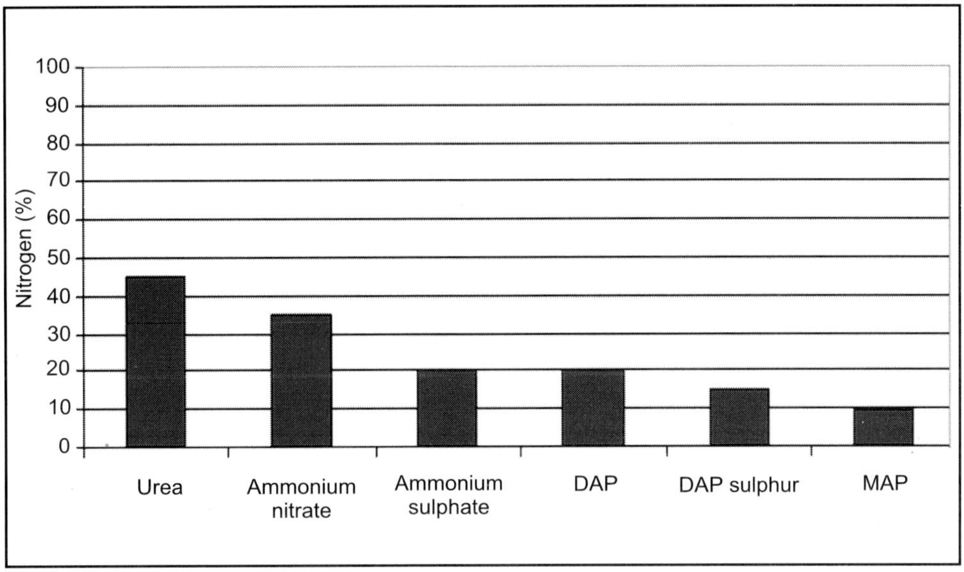

Fig. 2.4. Major nitrogen fertilizers.

urea, triple super phosphate, composite, potassium nitrate and sulfate, and Map fertilizers; ammonium nitrate and composite fertilizers comprise approximately 80 percent. During the last two years, the import of fertilizers has increased tremendously, and these include mainly nitrogenous fertilizers. Farmers prefer calcium as a nitrogenous fertilizer, and for this reason the use of this fertilizer has increased by 13 percent. Ammonium nitrate and urea fertilizer applications have decreased by 10 and 5 percent, but the classical fertilizer consumption has increased lately by 3 percent and reached to 5.4 million tons (Topbas et al. 1998; Atilgan et al. 2007).

The chemical fertilizers are used very intensively especially during winter, which results in a serious pollution of water resources, particularly wells (Kaplan et al. 1999). The amount of fertilizer applied during 1997–2001 varied between 91–121 kg/h (Anonymous 2002). Nearly 90 percent of the greenhouse agricultural production is practised in the Mediterranean region, out of which 14 percent is glasshouses, 34.7 percent plastic covers, 11.6 percent high tunnel and 39.7 percent low tunnel origin (Anonymous 2001; Anonymous 2003; Atilgan et al. 2007). There is an excessive use of fertilizers in the greenhouses (Tuzel et al. 2005). The farmers use almost similar amounts of fertilizers unknowingly with the belief that more production will take place if excess fertilizer is used (Yilmaz et al. 2000), or simply because the owners in the neighbourhood apply the same amount. The data published by Atilgan et al. (2007) shows that 70 percent of the producers use 33 percent of the nitrogen fertilizer containing ammonium nitrate. This amount reaches up to 46 percent together with the use of other chemical fertilizers containing nitrogen. The natural organic fertilizers are also used in the area, which are rich in micro- and macronutrients. These provide nutrients and improve the soil structure as well. Some synthetic organic fertilizers (Ormin K, Biyofarm, NOF, Organica, Coplex, Orgevit, liquid and powder organic fertilizer, liquid and powdered chicken manure) are also used. The maximum levels of nitrogen used in the region are 10 times more. The situation for phosphorus, potassium, magnesium, iron, zinc, manganese, and copper is much worse (Kaplan et al. 2001). The danger increases if we consider the waste water effluents from the factories established to meet the increasing fertilizer demand. The wastewater effluents of fertilizer factories contain ammonium and nitrate nitrogen levels well above the limits specified in the regulations (Anonymous 2004).

NITRATES AND LIVING BEINGS

Ammonium nitrate fertilizers are more complex in their mode of action. The ammonium portion is immediately available in the rumen and is passed across the rumen wall leading to an increase in the blood ammonia levels very shortly after consumption. The nitrate fraction, accumulates in the rumen and is converted to nitrite before crossing the rumen wall, creating nitrate poisoning conditions. Toxicity due to nitrates can be either acute or chronic (Hill 1999; Bjorne et al. 2004). Acute toxicity is expressed when nitrite enters the bloodstream in large quantities and oxidizes the ferrous iron of hemoglobin to the ferric form producing methemoglobin, which cannot transport oxygen and results in methemoglobinemia (Szaleczky 2000; Fewtrell 2004). It has been reported that unless immediate relief is provided, a proportion of methemoglobin that reaches to approximately 70 percent or more of the total hemoglobin may be fatal. The methemoglobinemia cases are generally restricted to anemic adults who genetically have high methemoglobin levels in their blood, and the situation is visually prominent in infants under 3 months of age and is known as blue baby disease (Al-Redhaiman 2000). The reports published by some researchers reveal that nitrate, nitrite and some other compounds can get converted to nitrosamines in human and animal digestive systems, which could have carcinogenic effects (Atilgan et al. 2007).

The reports show that over 80 percent of the total nitrate intake in humans originates from vegetables and some part from the drinking water (Croen et al. 2001; Anjana et al. 2007 a,b,c). Although the risks have not been fully evaluated, several developed countries have established maximum permissible levels for nitrate concentration in vegetable food crops, and other countries are considering the establishment of similar standards. As a result of increasing pollution of sources adjacent to agricultural activities, the World Health Organization has set a limit of 10 mg of nitrate nitrogen per litre in domestic water supplies (Van Diest 1986; Pionke et al. 1990; McPharlin and Aylam 1995). Research has shown that high concentrations of nitrate in plant food can decrease the level of ascorbic acid, a known inhibitor of the formation of carcinogenic compounds that contain nitrite.

There are acute and chronic effects of nitrate poisoning in animals (Cash et al. 2004). When acute poisoning occurs, most animals die before any signs of toxicity are observed. Death usually occurs within a few hours of feeding (Al-Redhaiman 2000). The clinical signs of acute poisoning are increased heart rate, blue-grey mucous membranes, excess saliva and tear production, weakness, depression, laboured or violent breathing, staggered gait, frequent urination, vomiting, muscle tremors, low body temperature, disorientation and an inability to get up. Animals are found in a lying position after a short struggle.

Chronic nitrate toxicity is a form of nitrate poisoning where the clinical signs of the disease are not observed. It is more common to see a reduction in the rate of weight gain, lower milk production, depressed appetite, and a greater susceptibility to infections. These production-related problems or losses are not often recognized and will occur when nitrate levels are at 0.5 to 1.0 per cent of the feed consumed (on a dry weight basis). This type of poisoning can cause abortions within the first 100 days of pregnancy by interfering with the implantation of the egg in the uterus. Calves affected by nitrate poisoning in the last 3 months of gestation are usually born 1 to 4 weeks premature, and most appear normal but die within 18 to 24 hours of birth. Newborn calves that survive, but are affected by nitrate poisoning, may have convulsions and seizures. Nitrates reduce thyroid gland activity and therefore, the metabolic rate of the animal, causing a depression in appetite and lower weight gains. Animals exhibiting signs of distress from chronic toxicosis may appear uncoordinated, belligerent, or have uncoordinated hind quarter movement, together with diarrhea, and eye lesions (Al-Redhaiman 2000; Gupta 2006).

In the Mediterranean region the greenhouse agriculture is intense, at many places drinking water pipelines are lacking and the wells in the vicinity are used by the labourers working in the area as well as by local inhabitants. A large number of these greenhouses lack drainage systems. The excess irrigation water thus moves into deeper parts of the soils and gets mixed up either with the groundwater or reaches the wells lying in the vicinity of the greenhouses or farms, thus polluting not only the waters but also producing adverse effects on soil flora. In some cases the nitrate concentration is (45 ppm) above the levels recommended by WHO (Kaplan et al. 1999). Nitrate contaminated well water, consumed along with certain feeds that are high in nitrate, can pose health hazards to domestic animals. These waters may cause livestock losses. Occasionally, when water contains insufficient nitrate to be toxic by itself, it may add to the toxicity of nitrate-containing feeds. Nitrite in water is 10 times more toxic to animals compared to nitrate in feeds.

The critical toxic level of nitrate varies according to the animal group (Smith and Guthrie 2004). Molybdenum, copper, iron, magnesium and trace elements such as manganese, are effective in the nitrate to ammonium conversion. Therefore, in animals are feed with plant material lacking these elements, toxic effects of nitrite occur in higher levels. The animals fed on a single type hay or silage with high nitrate content are most likely to get poisoned. Depending on the nitrate content of plants in the grasslands, poisoning can be seen in grazing animals. The symptoms start within 1-2 hours after the consumption of high nitrate feed; negative effects of mid-level nitrate feed can be seen several days later after intake. When methemoglobin level exceeds 20 percent, brown or bluish color appears on mucosa membrane in the mouth and tongue. Other symptoms could be seen when the levels are between 30–50 percent. The animal dies when the level reaches to 70–80 percent (Smith and Guthrie 2004; Weiss 2004).

According to Oruc and Ceylan (2001a, b), the amount of nitrate in dry matter which could cause acute food poisoning is 2260 ppm. Animals suffering from nitrate poisoning should be treated with diluted 2 g methyl blue per 500 kg weight. It reduces the amount of methemoglobin providing more opportunity for oxygen transfer (Weiss 2004). It is important that before the treatment, identity as well as the density of the poisonous plant should be clear and physiological responses together with the symptoms in animals need to be well evaluated. The necessary interventions must be made by removing the animals from the area.

Chronic and acute effects of high nitrate containing feed should be taken into consideration carefully while feeding the animals. According to the data published by World Health Organization (WHO), daily nitrate ingestion limits should not exceed 5 mg per kg, and nitrite 0.4 mg per kg for adults (Bloom and Lampe 1983). In Turkey, the maximum acceptable levels for animal feed have been recommended as 15 mg/kg sodium nitrite (Reinink 1991; Reinink et al. 1994). On the other hand in 1995, the Scientific Committee for Food in the European Commission has announced the acceptable daily intake of nitrate as 3.65 mg kg^{-1} body weight (equivalent to 219 mg/day for a person weighing 60 kg) (SCF 1995; EC 2005), whereas the Joint Expert Committee of FAO, UN and WHO has established the acceptable daily intake of nitrate as 0–3.7 mg kg^{-1} body weight (Speijers 1996). On November 8, 2005, the European Commission adopted EC Regulation No. 1822/2005 and set the harmonized maximum levels for nitrate in spinach, baby foods and processed cereal-based foods (Seginer 2003). The limits given vary according to the season, with higher nitrate levels permitted in winter-grown vegetables. The Environmental Protection Agency (EPA) of America has established the reference dose for nitrate as 1.6 mg nitrate-N kg^{-1} body weight per day (equivalent to about 7.0 mg NO_3 kg^{-1} body weight per day) (Mensinga et al. 2003). In view of this, for a real assessment, nitrate content in all sources as well as their average daily consumption amount should be considered before making a decision.

NITRATE AND PLANTS

Plants use particular portion of the applied fertilizers and the rest ends up in aquifers threatening plant, animal and human health (Bruning-Fann and Kaneene 1993; Brown et al. 1993; Eichholzer and Gutzwiller 1998; Majumdar and Gupta 2000; Croen et al. 2001). Various plants accumulate nitrates if certain conditions exist (Blom-Zandstra 1989; Durner and Klessig 1999; Chen et al. 2004). Some of the more common accumulators are commerical crops, weeds, greenfeed barley, rye, oat and wheat, beet tops, sorghum, millet, sunflower, flax, thistles, mustards, canola plants, kochia, nightshade, smartweed, and white ragweed (Gastal and Lemaire 2002). Critical level of nitrate varies between 2000–2500 ppm in forage crops, which is related to drinking water. Cereal crops are usually planted into well fertilized, manured or recently plowed/broken grassland or pasture. These crops are also harvested at an early stage of development (milk to dough stage) when they contain the highest nitrate content. Some plants usually do not require soil available nitrogen due to the presence of nodules on their roots, which absorb gaseous nitrogen from the atmosphere and convert it into ammonia and subsequently other forms of plant nitrogen with the help of *Rhizobium* bacteria located in the nodules through nitrogen fixation contributes up to 85 percent of the nitrogen content in plants.

Studies on the effect of nitrogen rate and its comparison with the production of plants has revealed that after applying different amounts of nitrogenous fertilizers to the plants in a farmland, an increase is observed in the plant growth at 50 kg N/ha as compared to 10 and 25 kg, but overapplication of nitrogen, 75 and 100 kg adversely affects the plant growth, depression in growth being more prominent at 100 as compared to 75 kg, through a decline in the final dry matter production. The cumulative adverse effects become more significant following applications for a few years. This stresses the fact that low to moderate rates of nitrogen should be used to strategically increase plant production on the farmlands. In addition, the component of the nitrogen-fixing species and seasonal effects of the application on a farmland are important in the nitrogen nutrition of the plants used for feeding animals (Fig. 2.5). Usually, economic responses to nitrogen fertilizers can be obtained following application in late winter, provided there are

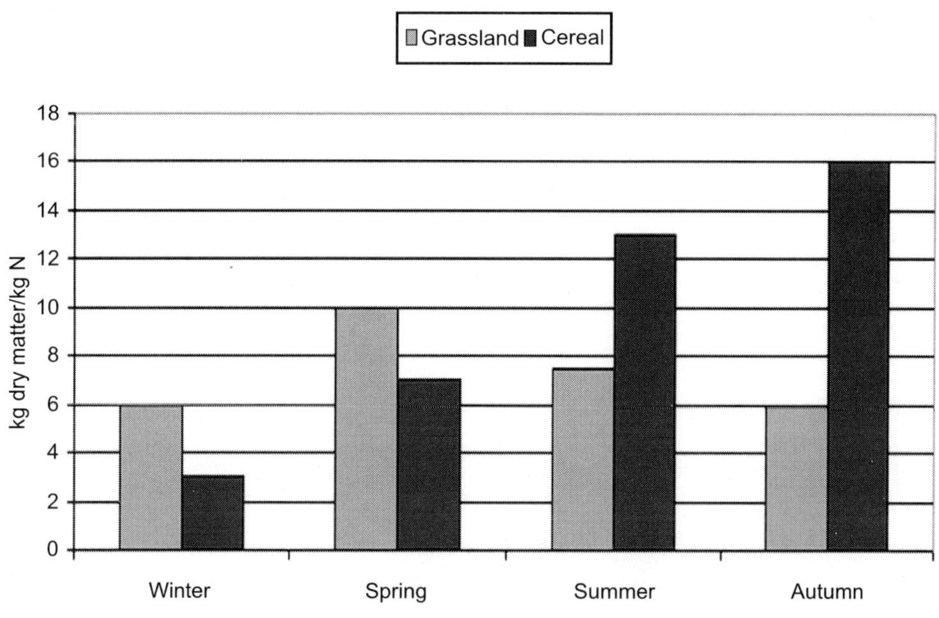

Fig. 2.5. Responses of plants after applying 50 kg N/ha to two farmlands in different seasons.

sufficient other companion species that can grow well at that time of year. Generally, fertilizers change the plant composition of farmlands, and plant response to nitrogen is best when it limits their potential to grow and the level of response depends largely on its deficiency at that time. Sometimes responses which are not economic can also occur when fertilizers are applied during autumn-early winter. Lower nitrogen application rates reduce potential changes in the composition of farmlands because some plants are stimulated less than with heavier applications.

The levels of nitrogen in the farmland species get affected by the rotation length and rate of fertilizer applied. Rotational grazing is widely practised on irrigated lands. The use of nitrogen is well suited to this form of farm management because resting between grazings gives time to respond to the applied fertilizer before it is grazed again. The length of a pasture rotation affects the quantity of pasture grown within each rotation, but also can change the relative response to nitrogen (Fig. 2.6). When rotation on the farmland is short, there is loss of response to nitrogen fertilizer and in the case of marginally responding plants, this affects the economic application of the fertilizer. Although N rate increased the N content of the pasture, the effectiveness of the fertilizer in achieving this declined with increased rotation length. Generally, if weather conditions are favorable for plant growth, plants take nearly a month from the time of application to the next grazing, to respond adequately. Under less suitable growth conditions, it will take longer for most of the response to occur. Therefore, the rotation needs to be extended by an appropriate amount. Consumption by stock of relatively small amounts of N fertilizer granules can be toxic to the stock. Dissolving the nitrogen fertilizer granules by irrigation or rainfall before allowing stock access to the pasture should

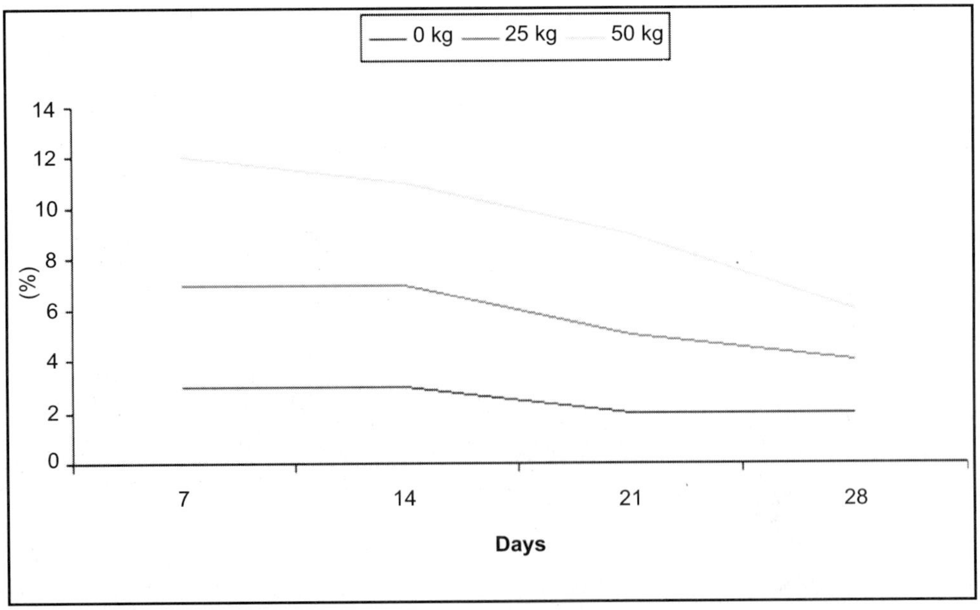

Fig. 2.6. Effect of nitrogen rate and rotation length on plant nitrogen content.

reduce any potential problems with stock consuming fertilizer on the pasture or from spillage from a spreader.

Several studies have indicated that plant responses could be much greater when nitrogen is applied within seven days after grazing (Fig. 2.7). The timing of the application of nitrogen is probably more critical with lower rates of fertilizer applied.

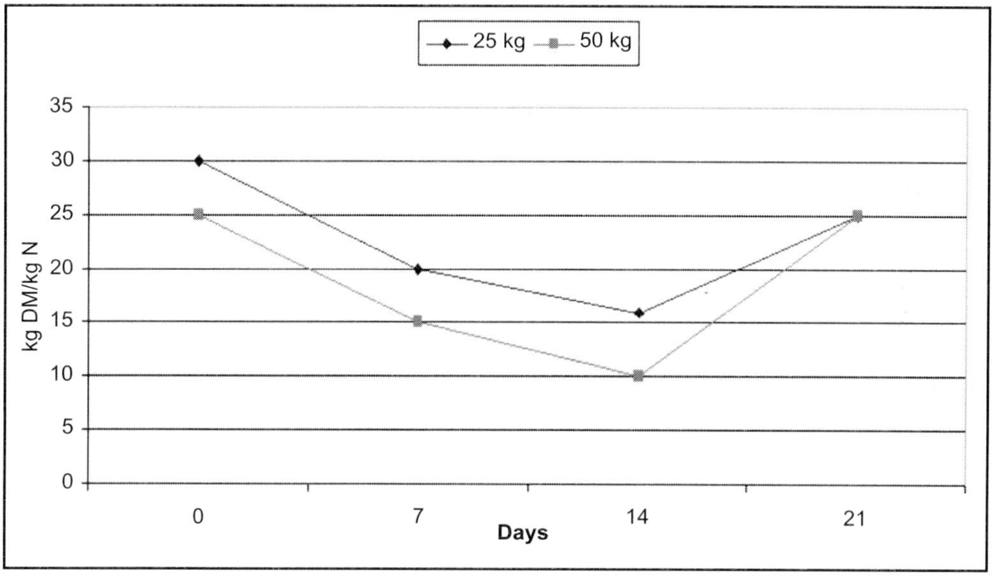

Fig. 2.7. Cumulative response of plants after single application of nitrogen at different times.

Plant growth varies from day-to-day and week-to-week throughout the season. The frequency of nitrogen application to plants greatly influences the production, but not the response. Plant responses with less frequent applications will fluctuate depending on nitrogen application. However, residual effects of the fertilizer on plant growth are often small in comparison with the initial response obtained after application.

The repeated use of nitrogen to increase the production within a short period reduces legume growth, unless the amount of nitrogen applied during this period is restricted. In the case of fields with a mixture of grass and legume, 50 kg N/ha applied over a season is not effective in the production of legumes, but if 100 kg N/ha is applied, total legume production is significantly reduced. The application of 50 and 100 kg N/ha over a season to a cereal-legume field has no significant effect on total legume production (Fig. 2.8). At the same time total dry matter production in such fields (Fig. 2.9) shows that the effect of intensive nitrogen use in the cereal-legume field is related to the quality and possibly milk production due to relatively low digestible cereal domination in the field.

Nitrogen fertilizers are used widely in banana plantations in our Mediterranean region. Its use over a number of times in a year to maintain a higher rate of plant growth and with the expectation that it will fill the feed gaps may limit the feed

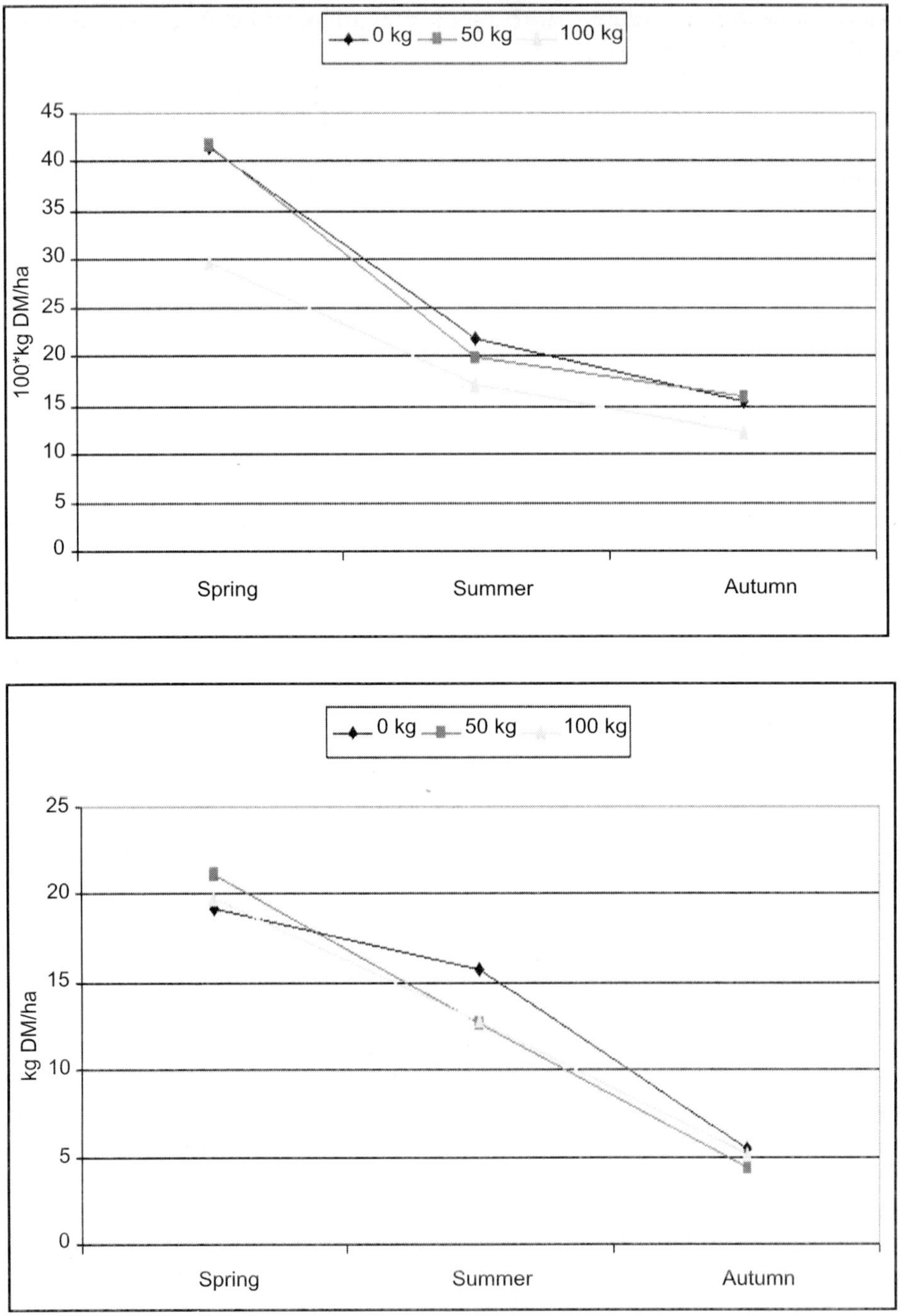

Fig. 2.8. Total legume production in two fields with 0, 50, and 100 kg N/ha applied twice in a season.

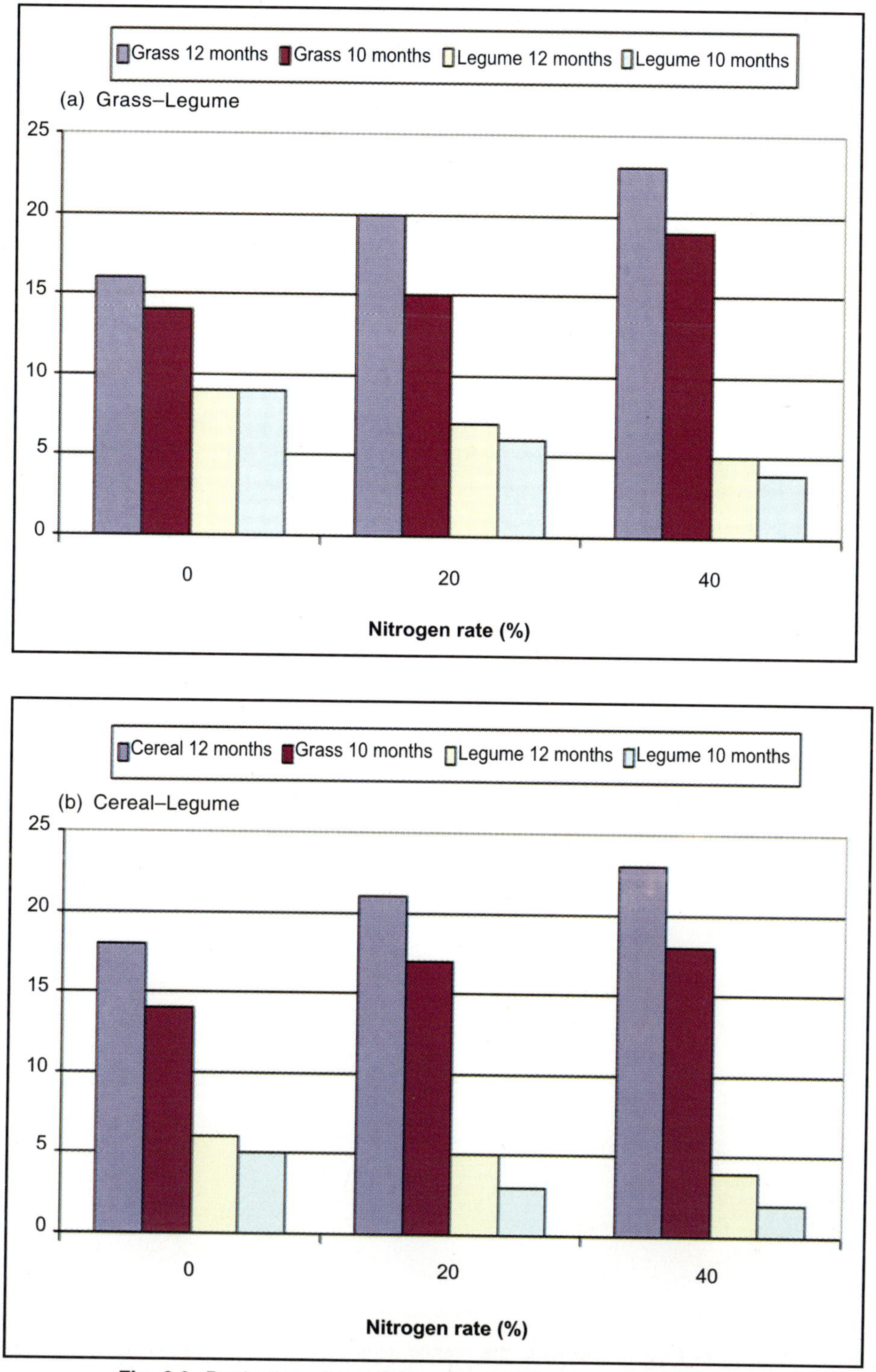

Fig. 2.9. Production of two fields after applying 0 or 20 or 40 kg N/ha.

availability for prolonged periods. The consequences of repeated applications need to be understood if its use is to remain economical.

An excess use of nitrogen fertilizers results in the accumulation of potentially harmful nitrate concentrations especially in the leaves of edible plants like spinach, lettuce, cabbage, etc. (Lorenz 1987; Alt and Fuller 1988; Luo et al. 2006; Anjana et al. 2007 a,b,c; Anjana et al. 2009 a,b). In general, the harmful effects originating from an overuse of nitrogenous fertilizers in the plants are bad quality of fruits (thick peel, spongy, not resistant) (Fig. 2.10), decrease in the resistance to drought, cold, diseases and pests, soft tissue and cell walls, fast growth followed by lying of whole plant and thus lower grain yield, but increased shoot growth, and finally monetary loss. The spinach and chenopodium leaves contain nitrate in concentrations higher than the recommended daily intake limits (Anjana et al. 2007 a,b,c). (Roorda van Eysinga 1984; Dich et al. 1996; Turkdogan et al. 2003; Anjana et al. 2007 a,b,c). Generally, healthy plants often contain 3–4 percent nitrogen in the shoots, which is a higher concentration than many other nutrients. The important winter crops like celery are grown on the soils rich in nitrogen. A supplemental fertilization is essential for optimum production. Nitrogen fertilizers are seldom applied because soil oxidation generally supplies celery's requirements. There are situations in which nitrogen applications can be reduced or eliminated when a crop follows nitrogen fixing plants. Application of nitrogenous fertilizers used in the dairy farms is very important to increase the plant production and help reduce expected feed gaps. Therefore, strategic nitrogen fertilizer use is defined as its addition to increase plant growth. The benefits of applying these fertilisers are best with responsive plants, but at present no soil test is available for predicting the response of such fertilizers. However, indicators of responsive plants are obvious urine patches after pasture topping, pale green colour in the plants (Fig. 2.11), and decrease in the efficiency with increasing rates of fertilizer application.

Fig. 2.10. Nitrogen related developments in the orange symptoms in fruits (high nitrogen-left, optimum nitrogen-right).

Fig. 2.11. Nitrogen deficient corn leaves.

The plant species differ in the accumulation of nitrates (Aksoy et al. 1999). Under normal growth conditions, aleppo grass, *Sorghum vulgare, S. sudanense, Chenopodium album* and *Celosia argentea* usually contain high levels of nitrate, whereas in corn the level increases under arid conditions. Similarly, oats, rye, wheat, barley, triticale and grasses tend to accumulate nitrate if it is available in the soil at high levels. The nitrate level recorded in *Festuca pratensis* (1100 ppm) is 3 times higher than *Phleum pratense* under same environmental conditions (Ebert et al. 1997). The nitrate content increases in forage crops when nitrogen containing fertilizers are applied, but the values vary according to the plant species (Sulak and Aydin 2005). In *Phleum pratense* and *Dactylis glomerata,* accumulation was 250 and 500 ppm respectively after applying 5.6 kg/hectare of nitrogen, but the values go up to 3600 and 17,000 ppm nitrate respectively if 16.8 kg of nitrogen/hectare are applied to the same plants (Singer 2002). In corn the amount of nitrate during three development stages was determined as 5400, 3600 and 600 ppm respectively. In general, the legumes accumulate less nitrate compared to grasses. Therefore, legumes can be planted in pastures as a precautionary measure (Smith and Guthrie 2004).

Nitrate content of plants is moderate in the leaves as compared to the stem and roots, whereas in high-energy grains it is rather low. The form of nitrate fertilizer applied is effective on the nitrate content of the plants. The results published by Aktas et al. (1993) show that in corn accumulation of nitrate was higher after an application of ammonium sulfate nitrate as compared to ammonium nitrate. Generally, the nitrate content in plants decreases with maturity, however, fodders containing high levels of nitrate do not show a decrease in the level of nitrate (Seay 1996; Smith and Guthrie 2004). Ahmed et al. (2000) have shown a reduction in nitrate content due to increasing phosphorus fertilization. The reason being growth reduction associated with limiting phosphate, which results in an increase in the root-shoot ratio, dry matter content, concentrations of sugars and organic acids, and reduction in the concentration of nitrate in the shoots (Buwalda and Warmenhoven 1999). Nitrate accumulation has also been studied in relation to some other chemicals. Foliar application of salicylic acid, molybdenum fertilizers and nitrification inhibitors and calcium can materially reduce the nitrate content of plants (Tzung et al. 1995; Ahmed et al. 2000; Zhou et al. 2000; Xu et al. 2005). In lettuce leaves, sugar levels and free amino acid concentrations increase under high supply of calcium and replace nitrate in the vacuole of the lettuce cell (Ahmed, 1996). There exists a negative correlation between the nitrate content and sulfate content in the lettuce plant (Blom- Zandstra and Lampe 1985) and, therefore, sulfur deficiency might lead to an increase in nitrate content. The nitrate levels in most crops (cabbage, carrot, beet, tomato, onion and potato) have been reported to decrease significantly under the influence of humic acid as well as zeolite (Cardenas-Navarro et al. 1998, 1999; Nazaryuk et al. 2002). Accumulation of salt in the soils can alleviate nitrate accumulation in vegetable crops to high fertilization and poor water supply, the water potential of the soil may become so negative that salination takes place and plants take up and accumulate nitrate as an osmoticum to adapt to water conditions. In such a situation, potassium compensates for the increased nitrate in maintaining the electroneutrality of the plant (Ruiz and Romero 2002; Chung et al. 2005; Gairola et al. 2009).

The amount of nitrate used by the plant also depends on its metabolic activity level (Inal and Tarakcioglu 2001; Stitt et al. 2002). Nitrates combine with carbohydrates during metabolic processes to form the amino acids glutamine and asparagine. These two amino acids are the base units for the formation of all other plant amino acids and subsequent protein. Nitrate that is brought into the plant accumulates in the stems and leaves when the glutamine and asparagine pathways are saturated. If the conditions are not optimal, the conversion rate of nitrate into amino acids is impaired. The plant usually uses up all available plant sugars, and this results in the plant accumulating soluble nitrogen as nitrate and ammonia.

The nitrate accumulation by plants in toxic amounts on soil high in nitrates is an exception rather than a rule. The factors responsible for nitrate accumulation in plants are mainly nutritional, environmental and physiological. Fertilizing with nitrogen or plowing under legumes can contribute to the nitrate content of the soil. The nitrate-utilization pathway depends on an adequate water supply, energy from sunlight and warm temperatures (above 13°C). Nitrogen fertilization and light intensity, in particular the diurnal changes, have been identified as the major factors together with many nutrients that influence the nitrate accumulation in plants. It varies in various parts of a plant and their physiological age (Delhon et al. 1995; Gaudreau et al. 1995; Santamaria et al. 1999; Anjana et al. 2007 a,b,c; Santamaria 2006). Excess nitrate also accumulates in drought stressed plants. The reason being that dry weather usually strikes at a time when nitrogen is being rapidly absorbed by the plant, and drought thus causes a reduction in the conversion of nitrates to protein due to water stress, therefore resulting in nitrate accumulation. If any of these conditions appears in an agricultural area, the potential for nitrate poisoning exists within a few days of harvest or grazing.

Generally, 85 percent of the nitrogen from urea is recovered from initially moist and wet soils within 4 weeks after its application to the fields with dry soil at the end of autumn season under different initial soil water contents and under standard flood irrigation. The lowest recovery is around 70, which occurrs on the initially dry soils; only 10 mm of rainfall is sufficient to improve the recovery. As such, application of urea should be done just before the expected rainfall or at places where irrigation facilities are avilable. The application of nitrogen fertilizers to the dry soil without irrigation or follow-up rainfall will produce negative effects on plant growth. Similarly, ammonium sulphate and DAP get easily lost from alkaline or heavily limed acid soils. Ammonium nitrate and MAP are more suitable alternatives for such soils due to less possibility to form the unstable ammonium carbonate, from which ammonia is lost. The losses of nitrogen through ammonia volatilization can also occur from DAP particularly on acid soils, but these losses are around 35 percent. Application of urea in cooler months results in a loss of 25 percent of the nitrogen to the atmosphere by ammonia volatilization. Overnight dews are possibly sufficient to dissolve urea, but are not sufficient to wash urea into the soil.

The effect of climate on nitrate accumulation has been studied at length by several workers, notable among these being Chadjaa et al. 2001 and Grzebelus and Baranski 2001. Gruda (2005) published a detailed review on the effects of different environmental factors on the greenhouse vegetables. Generally, low levels of nitrate have been recorded during high rainfall, but an accumulation occurs in warm and wet years, regardless of whether the nitrogen originates from organic or mineral sources. Nitrate levels are influenced by weather conditions more significantly than by the form and application rates of fertilizers (Custic et al. 2003). The accumulation varies with the season being higher in autumn-winter than in spring (Vieira et al. 1998; Santamaria et al. 1997, 2001). The reason is that in winter plants cannot use all the nitrogen available in the soil due to less favourable light and temperature conditions.

Aridity or water stress increases the level of nitrate in plants. The nitrate accumulation does not occur in fast developing plants; it occurs in the plants facing water scarcity, needed for normal growth and development, because decrease in nitrite to ammonium protein synthesis is blocked. After 3–7 days of a rain following a dry period, the level of nitrate in plants is elevated to the highest level. Similarly, when temperature decreases, low temperature and freezing reduce protein synthesis by blocking the plant growth and development, and thus nitrogen starts accumulating as nitrate within the plants. The application of herbicides like 2-4 D to the fields results in a temporary increase in the nitrate content of plants. If the application is stopped and high nitrate accumulating exotics are removed, the nitrate content in the plants decreases. Cloudy and foggy weather also stimulates nitrate accumulation in plants by interferring with the normal photosynthetic behaviour.

In agricultural practices, the biggest problem in the Mediterranean region is that with time farmers start increasing the amount of dose applied (Anonymous 2002). The totally dissolved fertilizers are used in the greenhouses by drip irrigation systems. The chemical fertilizers are used intensively especially in greenhouses during the winter months. All this is done usually by visually examining the plant or soil status. This leads to an inefficient use of water, fertilizer and energy, and ends up with environmental pollution, problems in the groundwater and drainage followed by a decrease in the yield and quality. Cleaning of a polluted environment is very difficult, expensive, and even in some cases impossible (Yilmaz et al. 2000; Yilmaz 2005). In 50 percent of the well waters, nitrate levels have been recorded above the limits accepted by WHO (45 mg NO^{-3}/L), and in some cases it is 3 times above the limit (165 mg NO^{-3}/L) (Kaplan et al. 1999). The only origin for such high levels can be attributed to the excessive use of fertilizers, both on the field as well as in the greenhouses. Although some of the producers use fertilizers upto 50 kg/da or even less which is in line with European standards, the levels of 51–75 kg/da, used in general are ten times higher than the averages recommended for the Mediterranean region. The average fertilizer use in Turkey is 91 kg/ha; the nitrate levels above 250 mg/kg and nitrite above 5 mg/kg become harmful (Gür et al. 1996; Anonymous 2002). Using fertilizers withought analyzing soils hinders their optimal use, increases the costs, and reduces the amount and quality of the product, as well as pollutes the environment (Gok et al. 1998). The excessive use of fertilizers results in the flow of nitrogen and phosphorus from the surface soil and drainage waters to the water resources, which ultimately effects the drinking water reserves, and increases the nitrate concentrations in the leaves of fresh vegetables (Michal 1998; Karaman et al. 2000; Oruç and Ceylan 2001 a,b). The tomato farming in the area constitutes 76 percent of all enterprises. The producers prefer tomatoes because of economic gain and marketing is easy, with a strong return much easier than other products. The chemical fertilizer usage is more in tomato cultivation, and more than 37 percent of these enterprises use fertilizers to get more earnings. Apart from the organic fertilizers, they use 50 g/da or more (N + P_2O_5 + K_2O) chemical fertilizers. The questionnaire distributed among the growers revealed that about 91 percent of the growers are aware of organic fertilizers, but only 35 pecent use organic fertilizers in greenhouses, 19 percent use them "for plant nutrition, 15 percent because of recommendations from pesticide dealers, and 31 percent for the root development and land smoothing. Some use them simply to add organic matter to the soil and for soil improvement. Nearly 54 percent of the industries use organic manure for plant nutrition to increase the yield. Looking at the quantities of farm fertilizers used, the results show that 38 percent of the greenhouses use 2 tons, 6 percent 2–3 tons and 24 percent 3 tons or more. It was observed that 32 percent of the industries did not use farm manure and were not interested in explaining the amount used. Nearly 70 percent of the growers had not analyzed the soils before applying the fertilizers, 55 percent had applied the fertilizers according to the recommendations from dealers, 24 percent by past experience, and only 21 percent were analyzing the soils depending on the plant and air conditions. Out of these applications, 36 percent of the organic fertilizer used was obtained from dealers, 45 percent from the market and 19 percent from the livestock enterprises and villages. The producers were using 29 percent of the organic fertilizer as farmyard manure, 18 percent from the one sold at the market (Ormin K, Biyofarm, NOF, Organica, = oplex, Orgevit), 11 percent used liquid chicken manure, 12 percent liquid manure and 30 percent other liquid and solid commercial organic fertilizers. Nearly 87 percent of the total organic fertilizer used has been applied to the tomato growing and 13 percent for other products. Approximately 36 percent of the enterprises declared that they were using more than 100 kg/da from the commercialised fertilizers.

The average nitrate content in fresh vegetables cultivated in the greenhouses were 3.68, 3.40 and 2.86 g/kg for lettuce, parsley and radish respectively. In the vegetables grown on the farm, the values for lettuce, cabbage, cauliflower, spinach, parsley, carrot, radish were 1.56, 1.06, 0.42, 0.84, 0.84, 0.50, 1.53 g/kg respectively. The fresh vegetables like lettuce, cabbage, cauliflower, spinach, radish and carrot grown as organic farming had 1.19, 0.96, 0.35 , 0.97, 1.31, 0.20 g/kg of nitrate (Alcicek and Baslar 1995) (Fig. 2.12a,b).

(a) Lettuce

(b) Rice

Fig. 2.12. Nitrogen fertilizers applied to lettuce (a) and rice (b).

Nitrogen fertilizers are very important for rice production (Fig. 6.12b). The studies carried out show that in rice best dose is 15 kg/da nitrogen in the pure form (Anonim 1990., Sürek ve Beşer 1997b). The nitrogen fertilizer should be applied at least twice in batches—50 percent before sowing and at the time of seedling appearance, the other 50 percent 55 days after sowing. If three applications are given, these should follow as 1/3 before sowing, 1/3 at the start of vegetative multiplication, and 1/3 definitely at the start of formation of panicles (Anonim 1990). Overdose application results in an increase in vegetative multiplication, as a result differently ripened panicles are harvested, resulting in an increase in the panicle sterility. Sometimes rice grains are not full, and with the decrease in the temperatures, there is no production, increase in the laying down, disease appearance increases, in particular risk of burning increases and production decreases. A decrease in the production, grain diameter and protein content has been recorded in case of low application of nitrogen fertilizers. If optimum dose is not applied, production and quality decrease at the same time (Anonim 1991; Beşer 1997b; Sürek ve Beşer 1997b).

Among the fodder plants corn, British lawn grass, Italyan lawn grass, clover, rape, sugarbeet leaf and mixed meadow grass contained 8.1, 8.2, 10.3, 6.4, 14.9, 14.1, and 5.4. g/kg dry weight of nitrate (Alcicek and Baslar 1995). Green fodder plants containing 4400 mg/kg nitrate can be fed without problem, those having 4400–8800 mg/kg are tolerable but water should have no nitrate, however, in those containing 8800–15000 mg/kg, the amount in the balanced diet should not be beyond 50 percent (Santamaria et al. 1997, 2001). The fodder plants containing nitrate 15,000 mg/kg and above are highly toxic (Alcicek and Baslar 1995). The permissible limits of nitrate levels in water samples are 44 (safe for humans and animals), 44–88 (normal for animals), 88–176 (not good for animals), 176–440 (can be used in the absence of nitrates in the feedlot), and 440–880 (no toxical effect, but fodder uptake effected unfavorably) mg NO/I (Alcicek and Baslar 1995).

Cereals are very responsive to nitrogen, but overapplication can reduce the yields and quality. Their response depends on the soil type and the soil nitrogen supply status. The yields are normal when nitrogen fertilizer is applied at lower levels between 7–9 kg/ha or 7–8 kg/ha for wheat and barley respectively, but the yields could be increased by applying 35 kg/ha in wheat and 25 kg/ha in feeding barley. Depending on the soil type, weather conditions or management, the amount applied can be lowered to 20 kg/ha. For spring-sown malting barley, total nitrogen is applied to the seedbed. If it is grown after beet tops or turnips or a legume, the amount is reduced to 25 kg/ha. However, if there is heavy rainfall during the early growth period, additional 25 kg N/ha is applied (McNaeidhe 1985). In the case of winter wheat, the nitrogen requirement is higher than other cereals, but factors like sowing season as well as soil type affecting the yield potential of the crop also affect its nitrogen requirement (McNaeidhe 1985). In general, 75 kg N/ha should be the maximum level applied to avoid poor establishment in spring cereals.

The response of sugarbeet to nitrogen fertilizers varies between 10–20 percent; rest of the nitrogen required by the crop is provided by nitrogen reserves in soil which are said to lie around 15,000 kg/ha N in the root profile (60 cm depth). In the surface soils the amount available is around 15 percent, which is made available by biological processes to different extents over different time periods. However, the amount varies from field to field and is dependent on rainfall criteria and soil type (Herlihy and Hegarty 2001; Coulter 2001).

MOLECULAR ASPECTS OF NITRATE ACCUMULATION

The nitrate content varies markedly with the plant species, cultivars of the same species, the genotypes with different ploidy, and even in different parts of the same plant (Ourry et al. 1997; Grzebelus and Baranski 2001; Harada et al. 2003; Anjana et al. 2007 a,b,c). After seeding, much of the nitrate taken up by the plant is used for root and shoot development. During the initial growth, the major part is retained in the root, but as the growth proceeds the plant is able to convert more nitrate to amino acids. This change in shoot to root ratio improves the nitrogen balance. Nitrate accumulates in the plant as long as the roots supply more nitrate than the plant can use. The highest nitrate levels are found in the lower one-third of the stalk or stem. Concentrations in the leaves and flowers are lower and the seeds or grain kernels are usually nitrate free.

The shoot nitrate content is genetically determined and likely to be controlled by several genes (QTLs) (Harrison et al. 2004). The causal factors might include genetic differences among genotypes in the enzymes of nitrogen metabolic pathway, the uptake rate of nitrate as well as other elements needed for enzyme activity, or differences in generation of electron donors required in the assimilative pathway. The differing capacities of nitrate accumulation can also be correlated with differing locations of nitrate reductase activity (Andrews 1986), differences in photosynthetic capacity (Behr and Wiebe 1992), ability to generate and translocate respiratory substrate and reducing equivalents, or differences in capacity to translocate the absorbed nitrate to reduction sites. Nitrate accumulation decreases with increasing carbohydrate concentration in the vacuoles, being negatively correlated to sugar concentrations (Blom- Zandstra and Lampe 1985) and dry matter content (Reinink et al. 1994). In general, the genotypes with high dry matter content could have a high carbohydrate content in their vacuoles and thus need little nitrate to maintain their osmotic value (Reinink et al. 1994). The studies using mutants and transgenic plants have revealed a number of genes that can affect the concentration of nitrate in a plant, e.g., genes encoding nitrate reductase, a putative anion channel, CLC-a, glutamine synthetase, and ferredoxin-dependent glutamate synthase (Scheible et al. 1997a,b; Geelen et al. 2000; Harada et al. 2004). Loudet et al. (2003) identified eight quantitative trait loci (QTLs) for nitrate content on a dry matter basis, whereas Hirel et al. (2001) identified five quantitative trait loci for nitrate content in dry matter of maize, one of which included a gene encoding glutamine synthetase. Nitrate storage in the vacuoles is affected by multiple processes. Therefore, numerous gene products can potentially influence the naturally occurring variation in free nitrate levels at the whole tissue level (Harada et al. 2004). Harrison et al. (2004) have studied genotypic variability in shoot nitrate content and found that it was mainly due to an increase in the ion uptake regardless of biomass production. The positive correlation between the shoot nitrate content and the steady state level of mRNA encoding high-affinity nitrate transporters suggests that the higher nitrate flux is due to enhanced expression of transporters, indicating thus that nitrate transport is one of the main checkpoints controlling shoot nitrate accumulation. The nitrate content can therefore be reduced through breeding strategies without affecting the biomass production. It is necessary to take into account that the control of nitrate uptake and its accumulation in the plant may be subjected to genetic variability regardless of the plant's demand. This variability allows for an adaptive regulatory control mechanism depending on soil nitrate availability. Thus, the selection of genotypes that accumulate less nitrate may contribute significantly to reduction in the nitrate consumption by humans through vegetables and the subsequent risk of nitrate poisoning (Tohgi et al. 1998; Ustyugova et al. 2002). Another method of circumventing the problems associated with nitrate accumulation may be to breed cultivars that do not accumulate nitrate even under heavy nitrogen fertilization.

FUTURE PROSPECTS

The fertiliser applications are becoming a great threat by becoming major polluters as they enter our environment. People have started thinking about the future of nitrogenous fertilizers due to their negative effects on human and environmental health. The threat is extending from the local level to global scale, and from deep groundwaters to the stratosphere. Its presence in our waters in high amounts causes life-threatening "blue baby disease" in our children. Overapplication of nitrogenous fertilizers around the agricultural fields is a continuous danger for us through the wells and underground waters, which may have dangerous amounts of nitrate accumulation. Excess nitrogen reaches the lakes and the sea leading to eutrophication. The nitrogen fertilizers also increase the soil acidity by increasing the degree of dissolution of trace elements and can result in the accumulation of more trace elements in our drinking waters. It also increases the level of nitrogen oxides in the atmosphere, thus contributing in the destruction of ozone layer, as well as an increase in the greenhouse effect in the troposphere. Nitrogen oxides stay more than 100 years in the atmosphere, and each molecule of nitrogen oxide absorbs 200 times more radiation than carbon dioxide. This will also play an important role in global climate change. As a result, it could be concluded that the world has been caught to the desirability of nitrogen fertilizers very quickly, and faced with its bad consequences before thinking and doing research on the effects on environmental health and the possible influence on climate. As a remedy to this dangerous trend, two focal points should be considered: first, we should encourage organic farming, and reduce the use of artificial fertilizers and pesticides.

The surveys concerning nitrate content of fodder are extremely inadequate in the region. Nitrate-containing forages may be used with proper precautions. Problems with nitrate toxicity can be avoided with proper management of forage and livestock. Drought-stricken forages may contain toxic amounts of nitrates for cattle, and prevention of problems is essentially the only valid approach to avoiding economic loss (Balabanli et al. 2006). Suspected forages should be tested and results used in the feed management. Limiting forage intake, adapting cows to forages, mixing problem forages and nitrate-free feeds and preventing animals from eating nitrate-containing forages, especially when hungry, are important in avoiding cattle losses. Silages suspected of toxic levels of nitrate should be allowed to aerate overnight before feeding (Sund et al. 2004).

Silaging reduces the nitrate content of the fodder between 30–60 percent. For this reason, at least 3-week period should be allowed for silage fermentation. Ammonia, butyric acid and non-protein nitrogen levels of the silage increase when nitrates are present in the silage crop. Silages that contain high nitrates have a higher final pH. The highest nitrite levels occur when the silage pH is 4.9 to 5.7. The breakdown of nitrate during the ensiling fermentation is caused by nitrate reductase enzymes. *Enterobacteria* are mainly responsible for the degradation of nitrate in silage. Similarly, *Clostridia* can reduce nitrite to ammonia, but this is only achieved when pH levels are relatively high. *Lactobacilli* reduce nitrate and cause lactate to be reduced to acetic acid, thus raising the final pH level in the silage. The type of plant material being ensiled and management techniques used influences nitrate losses. Nitrate reduction occurs when silage pH is above 5. A temperature rise is expected during ensiling, but if temperatures remain over 40°C for more than a few days, a check for nitrates and nitrites should be done before feeding any of the silage to livestock. Nutritional value of forage crops, is closely related with the quality. The most important criterion in the quality of forage crops is crude protein in dry matter. There is a positive correlation between nitrogen fertilization of plants and their crude protein content (Rubio et al. 1996). However, the nitrogen fertilization of forage crops influences the level of toxicity in animals depending on the nitrate content of plants (Zialdi et al. 2000). 2500 ppm or more nitrate in dry matter content has been reported to be an acceptable criterion for the quality of herbs (Ebert et al. 1997).

There is an immediate need for considering the estimation of the amount of nitrogen fertilizers to be applied to the fields or under glasshouses as kg N/ha rather than the amount of fertilizer per ha. At the same time fertilizer application and irrigation practices reducing nitrogen losses in runoff should be used such as the fertilizer should not be spread near the bottom end of the irrigation bay, unfertilized strip should be left at the drainage end, irrigation should be stopped before it reaches the drain, drainage water should be reused, fertilizer should be spread after the runoff, and some free water should be left at the top to dissolve the fertilizer particles and allow the diffusion of the applied N into the soil.

The fertilizer dealers provide free consultancy services to the farmers, which turns into a trade. They should instead organize training workshops for the producers in order to increase their environmental awareness. The producers mainly use the fertilizers according to the amount of production; they do not think of any criteria apart from the yield and no attention is paid to the possible environmental problems. The importance of environmental education comes to the forefront here. With an increase in the level of education of the users, the importance given to the analysis increases. Such people avoid unnecessary and excessive fertilizer usage due to environmental awareness.

Major part of the soil nitrogen reserves occurs in complex organic forms, which become available as the organic matter content of the soil is broken down to release nitrate and ammonium forms of mineral nitrogen. The soil nitrogen supply status can be deduced from the previous cropping, and manurial history of the soil, and should be categorised into an index system.

For the usage of nitrate containing fodder, following recommendations can be made: feed containing 1000 ppm nitrate could be used securely; feeds with nitrates between 1000–2000 ppm are slightly risky; forages containing nitrate between 2000–3400 ppm should not be given to animals, if given the ratio limit should not exceed 50 percent; forages containing nitrate between 3400–4000 ppm should be kept away from the pregnant animals. If given the ratio limit should not exceed 25 percent; feeds containing more than 4000 ppm nitrate forage are poisonous and should not be given to the animals; high nitrate fodders must be given at 2–3 weeks intervals. The nitrate levels in the forage crops growing on dry and wet soils could be reduced by late harvesting. Animals should not consume high nitrate feed when they are hungry. Intravenous injection of methylene blue in saline solution (4%) for horses must be in the 1–2 mg/kg range, while cattle and sheep require 20 mg/kg to obtain satisfactory results. By using sudans or sudan-sudan hybrids instead of sorghum-sudan or sorgo-sudan hybrids, nitrate accumulation potential can be reduced to a great extent. Although impractical, in most situations, contaminated water supplies can be made safe by boiling.

Right amounts should be applied at the right time to prevent yield losses due to wrong application of fertilizers and methods. Qualified researchers should guide the producers about the use of fertilizers in vegetable farming in greenhouses. The producers require a serious training from government or private sector (Kaplan et al. 2001; Tuzel et al. 2005; Atilgan et al. 2007). With an increase in the level of training, the number of producers using organic fertilizers as well as number of producers approaching institutions for soil analysis will increase. Soil testing is being suggested more often as an alternative for nitrogen fertiliser applications. Testing soils for nitrogen has been a useful practice in the drier regions for many years, and in such regions fertilizer rates can be adjusted to account for NO_3^- found in the soil prior to planting. The producers should arrange the analysis of soil and plant samples in the field and greenhouses for a sustainable production and prevention of soil pollution through irrigation (Guo et al. 2006).

Easily accessible laboratories should be established in the region, which will analyze the samples at a reasonable price. Restrictions should be imposed on the producers in the use of different types of fertilizers in the irrigation water through drip irrigation in order to prevent the wrong fertilization. While preparing a fertilization program, help should be extended to the producers to have conscious use of fertilizers, which is economically sustainable and enviroment-friendly. The ammonium nitrogen and nitrate nitrogen amounts specified in the regulations are very high in the wastewaters of fertilizer producing factories, which should be controlled.

Effective means of minimizing the accumulation of nitrates in vegetables include identification of low nitrate accumulating genotypes, manipulation of fertilizer regimes, exploiting natural diurnal fluctuations in nitrate concentration, and use of nitrification inhibitors. Rates needed to achieve different yields with different crops vary by region, and such decisions are usually based on local recommendations and experience. Nitrate consumption through vegetables can be kept low by harvesting them at the proper time. Some crops should be harvested in the afternoon of a sunny day when nitrate concentration in the leaves is low. It has been found that the leaf nitrate concentration is at its lowest at 1:00 p.m., when the sunlight is maximum.

Nitrogen should be applied at a place where all plants have direct access to it. Moist soil conditions are necessary for nutrient uptake. Placement below the soil surface can increase nitrogen availability under dry conditions because roots are more likely to find nitrogen in moist soil with such placement. Anhydrous ammonia must be placed and sealed below the surface to eliminate direct volatilization losses of the gaseous ammonia.

The major mechanisms for nitrogen fertilizer loss are denitrification, leaching, and volatilization. Denitrification and leaching occur under very wet soil conditions, while volatilization is most common when soils are only moist and drying (Benckiser et al. 1995). Therefore wet and dry conditions shold be controlled before fertilizer application as well as before grazing.

Although an inadequate use of nitrogen fertilizers reduces the benefits to the farmers (Mengel et al. 2006), it provides huge benefits to the manufacturers when sold at large amounts, who never warn the users about the negative environmental impacts due to excessive usage. The most important problem which manufacturers are not interested in solving is the optimal use of fertilizers.

CONCLUSION

Nitrogen losses from soil to water and air have become a concern at global level. Attempts are being made to put forth some of the best management practices to be used for controlling nitrogen loss from the fields or glasshouses (Magette and Carton, 1997). The implementation of various EU Directives, particularly the Action Programme (AP) required by the Nitrate Directive focuses an attention on tillage farming as a source of nutrient emission or loss to the environment. It is important that tillage farmers become more aware of the potential of their systems to cause nutrient emissions that may result in damage to the environment. A major element of this is the application of fertilizers to crops at optimum rates. One of the most important EU Directive related to the water quality objectives is Nitrate Directive-91/676/EEC which is important and challenging.

The Nitrate Directive introduced in 1991 focuses on nitrate losses from agriculture to water. Its objective is "to reduce water pollution caused or induced by nitrates from agricultural sources and prevent further such pollution". The issue of nitrate in water is of concern because elevated levels of nitrate in drinking water are considered a danger to human health. In addition, nitrate in water, particularly estuarine waters, results in the growth of algae and plants that cause undesirable shifts in the ecosystem. The Directive requires designation of nitrate vulnerable zones and to develop a list of measures to be implemented by the growers that will achieve the Directive's objective. There will be limits on the organic nitrogen loading from animals, and nitrogen fertilizer and manure applications will be limited. It is important that tillage farmers consider the potential pathways of nutrient loss from their systems and the strategies available to control them.

The emphasis on nutrient emissions from agriculture has up to now been primarily on water. There are three gases of concern—ammonia, methane and nitrous oxide. Only nitrous oxide is of concern from the perspective of agroindustry. This gas may account for up to 15 percent of the total global warming potential. Its global warming potential is estimated to be more than 250 times that of carbon dioxide. Moreover, its lifetime residency in the atmosphere is estimated to be about 130 years. Nitrous oxide also leads to ozone depletion. Globally, anthropogenic activity accounts for 64 percent of the total nitrous oxide emissions of which agriculture accounts for 92 percent. Farmers will be required to proportionally reduce the emission of greenhouse gases in the coming years.

Emissions of ammonia and methane relate primarily to animal production systems. However, ammonia emission from the land spreading of manure is an issue where a farmer uses animal manure as a nutrient source for crop production. Agriculture accounts for a major part of ammonia emissions. The sources are housing, manure storage, manure application to land, and urine and dung deposition in the fields. Once in the atmosphere, gaseous ammonia reacts progressively with acidic compounds forming ammonium salts, e.g., ammonium sulfate and ammonium nitrate. It is this addition of nitrogen to natural ecosystems, which is of environmental concern both in terms of its implications for biodiversity and the acidification of soil and aquatic systems.

Ammonia is one of the three most important compounds that are responsible for acid rain, the others being oxides of sulfur and nitrogen. Acid precursors in the atmosphere can cross national borders, and emissions from one country can adversely affect another country's environment. Therefore, the responsibility of achieving emission reduction targets rests primarily with the agricultural sector.

For this purpose, correct nutrient application is very important. This can be achieved by avoiding excessive nitrogen application. The timing of nutrient application is necessary and nutrients should be applied keeping with the Code of Good Agriculture Practice. In the nutrient application only calibrated machinery should be used and all fertilizers should be applied with care, avoiding uncropped areas, hedges, ditches, watercourses and wildlife habitats. Ploughing of long-term agricultural lands should be avoided, as it will result in an increased potential for N mineralization. In order to avoid soil compaction, avoid damaging the soil structure with late harvesting of some crops because this will increase the potential for runoff and erosion.

REFERENCES

Abdalla, M.I., Abdalla, M.M.A., Abdel-Aal, S.A. and Farrag, I.A. (1998). Effect of nitrogen fertilization on the yield and quality of some salad crops. *Hortsci. Abst.* **33:** 2.

Ahmed, A.H.H. (1996). Physiological studies on tipburn and nitrate accumulation in lettuce plants. *J. Agric. Sci.* **21:** 3971-3994.

Ahmed, A.H.H., Khalil, M.K. and Farrag, A.M. (2000). Nitrate accumulation, growth, yield and chemical composition of Rocket *(Eruca vesicaria* subsp. *sativa)* plant as affected by NPK fertilization, kinetin and salicylic acid, *In:* Proceedings of ICEHM, Cairo University, Egypt, pp. 495-508.

Anjana, Umar, S. and Iqbal, M. (2007a). Nitrate accumulation in plants, factors affecting the process, and human health implications, *A review Agronom. Sust. Dev.* **26:** 45-57.

Anjana, Umar, S. and Iqbal, M. (2007b). Identification of characteristics affecting nitrogen utilization efficiencies in wheat cultivars. *Arch. Agronom. Soil Sci.* **53:** 459-472.

Anjana, Umar, S., Iqbal, M. and Abrol, Y.P. (2007c). Are nitrate concentrations in leafy vegetables within safe limits? *Curr. Sci.* **92:** 355-360.

Anjana, Umar, S. and Iqbal, M. (2009a). Effect of applied potassium in increasing the potential for nitrogen assimilation in spinach (*Spinacia oleracea* L.). e-ifc (IPI) 20:8-10 http://www.ipipotash.org/udocs/ifc_No_20_rf2.pdf.

Anjana, Umar, S. and Iqbal, M. (2009b). Factors responsible for nitrate accumulation: A review. Lichtfouse, E. et al. (Eds.). Sustainable Agriculture, doi: 10.1007/978-90-481-2666-8-33.

Aksoy, A., Turel, I., Arslan, B. and Dede, O. (1999). The nitrate and nitrite levels of the tubers of potato plant fertilized with different doses of nitrogen fertilizers. *Tr. J. Veterin Anim. Sci.* **23:** 461-465.

Aktas, M., Gunes, A. and Baltutar, N. (1993). Effects of various forms of nitrogen sources on nitrate and nitrite accumulation in maize. *Tr. J. Agric. For.* **17:** 931-937.

Alcicek, A. and Baslar, S. (1995). The results of overaccumulation of nitrate in plants and waters. *Ekoloji* **4:** 15-18.

Al-Redhaiman, K.N. (2000). Nitrate accumulation in plants and hazards to man and livestock health: a review. *J. King Saud Univ. Agric. Sci.* **2:** 143-156.

Alt, D. and Fuller, A.M. (1988). Control of nitrogen status of lettuce by nitrate analysis of sap. *Acta. Hort.* **102:** 23-27.

Andrews, M. (1986). The partitioning of nitrate assimilation between root and shoot of higher plants: mini review. *Plant Cell Environ.* **9:** 511-519.

Anonymous, (2001). Tarimsal Yapi (Uretim, Fiyat, Deger). DIE Yayinlari, Ankara.

Anonymous, (2002). Turkiye Istatistik Yilligi. T.C. Basbakanlik Devlet Istatistik Enstitüsü Yayin No: 2466, Ankara.

Anonymous, (2003). Antalya Yas Meyve Sebze Ihracatcilari Birligi, 2002-2003 Yili Turkiye ve Antalya Sera Alani Varligi. Antalya Ili Euro GAP Protokolu HACCP Eylem Plani, Antalya.

Anonymous, (2004). Türkiye Cevre Atlasi. Cevresel Etki Degerlendirmesi ve Planlama Genel Mudurlugu Cevre Envanteri Dairesi Baskanligi, Ankara.

Atilgan, A., Coskan, A., Saltuk, B. and Erkan, M. (2007). The level of chemical and organic fertilizer usage and potential environmental impacts in greenhouses in Antalya region. *Ekoloji* **15:** 37-47.

Bacon, P.E. (1995). Nitrogen fertilization in the environment. Marcel Dekker, New York.

Balabanli, C., Albayrak, S., Turk, M. and Yuksel, O. (2006). Some toxic plants growing in rangelands of Turkey and their effects on animals. Suleyman Demirel University, Fores *Faculty J.* **A2:** 89-96.

Behr, U. and Wiebe, H.J. (1992). Relation between photosynthesis and nitrate content of lettuce cultivars. *Sci. Hortic.* **49:** 175-179.

Benckiser, G., Lorch, H.J. and Ottow, J.C.G. (1995). Quantification of total denitrification losses from undisturbed field soils by the acetylene inhibition technique. *In:* Alef, K. and Nannipieri, P. (Eds.). Methods in Applied Soil Microbiology and Biochemistry, Academic Press Ltd., London, 473-478.

Bjorne, H., Petersson, J., Phillipson, M., Weitzberg, E., Holm, L. and Lundberg, J.O. (2004). Nitrate in saliva increases gastric mucosal blood flow and mucus thickness. *J. Clin. Invest.* **113:** 106-114.

Blom-Zandstra, M. and Lampe, J.E.M. (1985). The role of nitrate in the osmoregulation of lettuce (*lactuca sativa* l.) grown at different light intensities. *J. Exp. Bot.* **36:** 2443-2452.

Blom-Zandstra, M. (1989). Nitrate accumulation in vegetables and its relationship to quality. *Ann. Appl. Biol.* **115:** 553-561.

Brown, J.R., Christy, M. and Smith, G.S. (1993). Nitrate in soils and plants. *In:* Agricultural publication G9804, Missouri-Columbia.

Bruning-Fann, C.S. and Kaneene, J.B. (1993). The effects of nitrate, nitrite, and N-nitroso compounds on animal health. *Vet. Hum. Toxicol.* **35:** 237-253.

Buwalda, F. and Warmenhoven, M. (1999). Growth-limiting phosphate nutrition suppresses nitrate accumulation in greenhouse lettuce. *J. Exp. Bot.* **50:** 813-821.

Cardenas-Navarro, R., Adamowicz, S. and Robin, P. (1998). Diurnal nitrate uptake in young tomato (*Lycopersicon esculentum* Mill.) plants: Test of a feedback based model. *J. Exp. Bot.* **49:** 721-730.

Cardenas-Navarro, R., Adamowicz, S. and Robin, P. (1999). Nitrate accumulation in plants: a role for water. *J. Exp. Bot.* **50:** 613-624.

Cash, D., Funston, R., King, M. and Wichman, D. (2004). Nitrate Toxicity of Montana Forages www.animalrangeextension.montana.edu/articles/forage/general/nitrate-tox.htm Moccasin.

Chadjaa, H., Vezina, L-P., Dorais, M. and Gosselin. A. (2001). Effects of lighting on the growth, quality and primary nitrogen assimilation of greenhouse lettuce (*Lactuca sativa* L.). *Acta Hortic.* **559:** 325-331.

Chen, B-M., Wang, Z-H., Li, S-X., Wang, G-X., Song, H-X. and Wang, X-N. (2004). Effects of nitrate supply on plant growth, nitrate accumulation, metabolic nitrate concentration and nitrate reductase activity in three leafy vegetables. *Plant Sci.* **167:** 635-643.

Coulter, B.S. (2001). Nutrient and trace element advice for grassland and tillage crops. Teagasc, Dublin.

Chung, J-B., Jin, S-J. and Cho, H-J. (2005). Low water potential in saline soils enhances nitrate accumulation of lettuce. *Commun. Soil Sci. Plant* **36:** 1773-1785.

Croen, L.A., Todoroff, K. and Shaw, G.M. (2001). Maternal exposure to nitrate from drinking water and diet and risk of neural tube defects. *Am. J. Epidemiol.* **153:** 325-331.

Custic, M., Poljak, M., Coga, L., Cosic, T., Toth, N. and Pecina, M. (2003). The influence of organic and mineral fertilization on nutrient status, nitrate accumulation, and yield of head chicory. *Plant Soil Environ.* **49:** 218-222.

Dapoigny, L., Tourdonnet, S.D., Roger-Estrade, J., Jeuffroy, M-H. and Fluery, A. (2000). Effect of nitrogen nutrition on growth and nitrate accumulation in lettuce (*Lactuca sativa* L.) under various conditions of radiation and temperature. *Agronomie* **20:** 843-855.

Delhon, P., Gojon, A., Tillard, P. and Passama, L. (1995). Diurnal regulation of nitrate uptake in soybean plants. I. Changes in NO_3 influx, efflux, and N utilization in the plant during the day/night cycle. *J. Exp. Bot.* **46:** 1585-1594.

Dich, J., Jivinen, R., Knekt, P. and Pentill, P.L. (1996). Dietary intakes of nitrate, nitrite and NDMA in the Finish Mobile Clinic Health Examination Survey. *Food Addit. Contam.* 13: 541-552.

Durner, J. and Klessig, D.F. (1999). Nitric oxide as a signal in plants. *Curr. Opin. Plant Biol.* **2:** 369-374.

Ebert, K., Koriath, H. and Rinno, G. (1997). The intensification of grass production through nitrogen fertilization and the nitrate problem. *XIII International Grassland Congress* **2:** 987-989.

EC (European Commission), Commission Regulation (EC) No. 1822/2005 of 8 November 2005 amending Regulation (EC) No. 466/2001 as regards nitrate in certain vegetables. *Official J. Eur. Union* **L293:** 11-13.

Eichholzer, M. and Gutzwiller, F. (1998). Dietary nitrates, nitrites, and AA Nitroso compounds and cancer risk: a review of the epidemiologic evidence. *Nutr. Rev.* **56:** 95-105.

Fewtrell, L. (2004). Drinking-water nitrate, methemoglobinemia, and global burden of disease: a discussion. *Environ. Health Persp.* **112:** 1371-1374.

Gairola, S., Umar, S. and Suryapani, S. (2009). Nitrate accumulation, growth and leaf quality of spinach beet (*Beta vulgaris* Linn.) as affected by NPK fertilization with special reference to potassium. *Indian J. Sci. Technol.* **2:** 35-40.

Gastal, F. and Lemaire, G. (2002). N uptake and distribution in crops: an agronomical and ecophysiological perspective. *Physiol. Plant* **53:** 789-799.

Gaudreau, L., Charbonneau, J., Venzina, L.P. and Gosselin, A. (1995). Effect of photoperiod and photosynthetic photon flux on nitrate in greenhouse-grown lettuce. *J. Plant Nutr.* **18:** 437-453.

Geelen, D., Lurin, C., Bouchez, D., Frachisse, J.M., Lelievre, F., Courtil, B., Barbier-Brygoo, H. and Maurel, C. (2000). Disruption of putative anion channel gene AtCLC-a in *Arabidopsis* suggests a role in the regulation of nitrate content. *Plant J.* **21:** 259-267.

Gok, M. and Ottow, J.C.G. (1998). Effect of cellulose and straw incorporation in soil on total denitrification and nitrogen immobilization at initially aerobic and permanent anaerobic conditions. *Biol. Fert. Soils* **5:** 317-322.

Gruda, N. (2005). Impact of environmental factors on the product quality of greenhouse vegetables for fresh consumption. *Crit. Rev. Plant Sci.* **24:** 227-247.

Grzebelus, D. and Baranski, R. (2001). Identification of accessions showing low nitrate accumulation in a germplasm collection of garden beet. *Acta Hortic.* **563:** 253-255.

Guler, S. (2004). Dunya'da ve Turkiye'de Gubre Tuketiminde Yasanan Gelismeler. *In:* Turkiye 3. Ulusal Gubre Kongresi, Karaman and Brohi (eds.), Tarim-Sanayi-Cevre, 11-13 Ekim 2004, Tokat, pp. 47-54.

Guleryuz, G., Gucel, S. and Ozturk, M. (2009). Nitrogen mineralization in a high altitude ecosystem in the Mediterranean phytogeographical region of Turkey. *J. Environ. Biol.* **31(2/3):** 10.

Guo, H., Li, G., Zhang, D., Zhang, X. and Lu, C. (2006). Effects of water table and fertilization management on nitrogen loading to groundwater. *Agric. Water Manag.* **82:** 86-98.

Gupta, S.K. (2006). Nitrate toxicity and human health. Proceedings of the Workshop on Nitrogen in Environment, Industry and Agriculture. New Delhi, India, pp. 8-10.

Gur, K., Zengin, M. and Uyanoz, R. (1996). Konya ve Cevresindeki Tarim Topraklarinda Yetistirilen Bugdaylarda Azotlu Gubrelerin Neden Oldugu Nitrat Kirliligi. *In:* Tarim-Cevre Iliskileri Sempozyumu, Dogal Kaynaklarin Surdurulebilir Kullanimi, 13-15 Mayis, Mersin, pp. 103-110.

Harada, H., Yoshimura, Y., Sunaða, Y., Hatanaka, T. and Sugita, S. (2003). Breeding of Italian ryegrass (*Lolium multiflorum* Lam.) for a low nitrate concentration by seedling test. *Euphytica* **129:** 201-209.

Harada, H., Kuromori, T., Horayama, T., Shinozaki, K. and Leigh, R.A. (2004). Quantitative trait loci analysis of nitrate storage in *Arabidopsis* leading to an investigation of the contribution of anion channel gene, AtCLC-c, to variation in nitrate levels. *J. Exp. Bot.* **55:** 2005-2014.

Harrison, J., Hirel, B. and Limani, A.M. (2004). Variation in nitrate uptake and assimilation between two ecotypes of *Lotus japonicus* and their recombinant inbred lines. *Physiol. Plant* **120:** 124-131.

Herlihy, M. and Hegarty, T. (2001). Prediction of nitrogen inputs for sugar beet. An evaluation of soil tests and soil management criteria in arable soils. Rural Environ Series No. 31, Teagasc, Dublin, p. 34.

Hill, M.J. (1999). Nitrate toxicity: myth or reality. *Brit. J. Nutr.* **81:** 343-344.

Hirel, B., Bertin, P. and Quillere, I. (2001). Towards a better understanding of the genetic and physiological basis for nitrogen use efficiency in maize. *Plant Physiol.* **125:** 1258-1270.

Inal, A. and Tarakcioglu, C. (2001). Effect of nitrogen forms on growth, nitrate accumulation, membrane permeability, and nitrogen use eficiency of hydroponically grown bunch onion under boron deficiency and toxicity. *J. Plant Nutr.* **24:** 1521-1534.

Kaplan, M., Sonmez, S. and Tokmak, S. (1999). Antalya-Kumluca yoresi kuyu sularinin nitrat icerikleri. *Tur. J. Agricu. Forest* **23:** 309-313.

Kaplan, M., Sonmez, S. and Alagoz, Z. (2001). Agricultural activity induced environmental pollution in the antalya region and solutions. *In:* Treatment 2000 Symposium and Exhibition, Ustun, B. (Ed.), 17-20 May, Istanbul, pp. 17-20.

Karaman, M.R., Brohi, A.R., Gunes, A., Inal, A. and Alpaslan, M. (2000). Yoresel degisik azotlu gubre uygulamalarinin Tokat bolgesinde yetistirilen bazi kislik sebzelerin nitrat akumulasyonuna etkisi. *Tur. J. Agri. Fore.* **24:** 1-9.

Lorenz, O.A. (1987). Potential nitrate levels in edible plant parts. *In:* Nielson, D.R. and McDonald, J.G. (Eds.). Nitrogen in the Environment, Vol. 2, New York: Academic Press.

Loudet, O., Chaillow, S., Krapp, A. and Daniel–Vedele, F. (2003). Quantitative trait loci analysis of water and anion contents in interaction with nitrogen availability in *Arabidopsis thaliana. Genetics* **163:** 711-722.

Luo, J., Sun, S., Jia, L., Chen, W. and Shen, Q. (2006). The mechanism of nitrate accumulation in pakchoi -*Brassica campestris* L. ssp. Chinensis (L.). *Plant Soil* **282:** 291-300.

Magette, W.L. and Carton, O.T. (1997). Agricultural pollution control. *In:* Environmental Engineering (Ed. Gerard Kiely) McGraw-Hill, 781-798.

Majumdar, D. and Gupta, N. (2000). Nitrate pollution of groundwater and associated human health disorders. *Indian J. Environ. Health* **42:** 28-39.

McNaeidhe, F.S. (1985). The effect of applying nitrogen at different times and rates on the yield of winter cereals on peatland. *Irish. J. Agric. Res.* **24:** 63-77.

McPharlin, I.R. and Aylam, P.M. (1995). Nitrogen requirements of lettuce under sprinkler irrigation and fertilization on spearwood sand. *J. Plant Nutr.* **18:** 219-241.

Mengel, K., Hutsch, B. and Kane, Y. (2006). Nitrogen fertilizer application rates on cereal crops according to available mineral and organic soil nitrogen. *Eur. J. Agronom.* **24:** 343-348.

Mensinga, T.T., Speijers, G.J.A. and Meulenbelt, J. (2003). Health implications of exposure to environmental nitrogenous compounds. *Toxicol. Rev.* **22:** 41-51.

Michal, F.D. (1998). A population based case control study on the association between nitrate in drinking water and non-Hodgkin's lymphoma. The Johns Hopkins University Press, USA, p. 285.

NAAS, (2005). Policy options for efficient nitrogen use. Policy paper No. 33, National Academy of Agricultural Sciences, New Delhi, pp. 1-4.

Nazaryuk, V.M., Klenova, M.L. and Kalimullina, F.R. (2002). Ecoagrochemical approaches to the problem of nitrate pollution in agroecosystems. *Russ. J. Ecol.* **33:** 392-397.

Nosengo, N. (2003). Fertilized to death. *Nature* **425:** 894-895.

Oruc, H.H. and Ceylan, S. (2001a). Bursa Yoresinde Sigirlarin Yemlerinde, Icme Sularinda ve Rumen Iceriginde Nitrat, Nitrit ve Kanda Methemoglobin Duzeylerinin Arastirilmasi. *J. Fac. Vet. Med.* **20:** 25-32.

Oruc, H.H. and Ceylan, S. (2001b). Bursa'da tuketilen bazi sebzelerde nitrat ve nitrit. Uludagi University, *J. Fac. Vet. Med.* **20:** 17-21.

Ourry, A., Gordon, A.J. and Macduff, J.H. (1997). Nitrogen uptake and assimilation in roots and root nodules. *In:* Foyer, C.H. and Quick, W.P. (Eds.). A molecular approach to primary metabolism in higher plants. Taylor and Francis, London, pp. 237-253.

Ozturk, M. and Secmen, O. (1995). Plant Ecology. Ege University Press, pp. 350.

Pionke, H.B., Sharma, M.L. and Hirschberg, K.J. (1990). Impact of irrigated horticulture on nitrate concentration in ground water. *Agric. Ecosys. Env.* **32:** 119-132.

Reinink, K. (1991). Genotype x environment interaction for nitrate concentration in lettuce. *Plant Breed* **107:** 39-49.

Reinink, K., Vannes, M. and Groenwold, R. (1994). Genetic variation for nitrate content between cultivars of endive (*Cichorium endiviae* L.). *Euphytica* **75:** 41-48.

Roorda van Eysinga, J.P.N.L. (1984). Nitrate and glasshouse vegetables. *Fert. Res.* **5:** 149-156.

Rubio, H.O., Wood, M.K., Gomez, A. and Reyes, G. (1996). Native forage quality, quantity, and profitability as affected by fertilization in Northern Mexico. *J. Range. Manag.* **49:** 315-319.

Ruiz, J.M. and Romero, L. (2002). Relationship between potassium fertilization and nitrate assimilation in leaves and fruits of cucumber *(Cucumis sativus)* plants. *Ann. Appl. Biol.* **140:** 241-245.

Santamaria, P., Elia, A. and Gonnella, M. (1997). Changes in nitrate accumulation and growth of endive plants during the light period as affected by nitrogen level and form. *J. Plant Nutr.* **20:** 1255-1266.

Santamaria, P., Elia, A., Serio, F. and Todaro, E. (1999). A survey of nitrate and oxalate content in retail fresh vegetables. *J. Sci. Food Agr.* **79:** 1882-1888.

Santamaria, P., Elia, A., Gonnella, M., Parente, A. and Serio, F. (2001). Ways of reducing rocket salad nitrate content. *Acta Hortic.* **548:** 529-537.

Santamaria, P. (2006). Nitrate in vegetables: toxicity, content, intake and EC regulation. *J. Sci. Food Agric.* **86:** 10-17.

Scheible, W-R., Lauerer, M., Schulze, E.D., Caboche, M. and Stitt, M. (1997a). Accumulation of nitrate in the shoot acts as a signal to regulate shoot-root allocation in tobacco. *Plant J.* **11:** 671-691.

Scheible, W-R., Gonzales-Fontes, A., Morcuende, R., Lauerer, M., Geiger, M., Glaab, J., Schulze, E-D. and Stitt, M. (1997b). Tobacco mutants with a decreased number of functional nia-genes compensate by modifying the diurnal regulation transcription, post-translational modification and turnover of nitrate reductase. *Planta* **203:** 305-319.

Scientific Committee on Food, (1995). Opinion on nitrate and nitrite. Annex 4 to Document 111/5611/95, European Commission (Ed.),. Brussels, p. 20.

Seay, R.L. (1996). Forage nitrate levels in greenfield bermudagrass as affected by nitrogen applications. Arkansas Agric. Exp. Station No. 450, 94-100.

Seginer, I. (2003). A dynamic model for nitrogen-stressed lettuce. *Ann. Bot.* **91:** 623-635.

Singer, J.W. (2002). Fresh versus field-cured grass quality, mineral, and nitrate concentration at different nitrogen rates. *Crop. Sci.* **42:** 1656-1661.

Smith, J.W. and Guthrie, L.D. (2004). Extension dairy scientists nitrate toxicity and prussic acid. poisoning in dairy cattle. www.ces.uga.edu/pubcd/b1153-w html.

Speijers, G.J.A. (1996). Nitrate. *In:* World Health Organization (Ed.), Toxicological evaluation of certain food additives and contaminants in food. Food Additive Series 35, Geneva, pp. 325-360.

Stitt, M., Muller, C., Matt, P., Gibon, Y., Carillo, P., Morcuende, R., Scheible, W-R. and Krapp, A. (2002). Steps towards an integrated view of nitrogen metabolism. *J. Exp. Bot.* **52:** 959-970.

Sulak, M. and Aydin, I. (2005). Nitrate accumulation in forage. *J. Fac. Agric. OMU,* **20:** 106-109.

Sund, J.M., Niedermeier, R.P. and Burris, R.H. (2004). Watch out for silage gas! http://www.cdc.gov/nasd/docs.pdf University of Wisconsin- Madison Extension.

Szaleczky, E., Pronai, L., Nakazawa, H. and Tulassay, Z. (2000). Evidence of *in vivo* peroxy nitrite formation in patients with colorectal carcinoma, higher plasma nitrate/nitrite levels, and lower protection against oxygen free radicals. *J. Clin. Gastroenterol.* **30:** 47-51.

Tohgi, H., Abe, T., Yamazaki, K., Murata, T., Isobe, C. and Ishizaki, E. (1998). The cerebrospinal fluid oxidized NO metabolites, nitrite and nitrate, in Alzheimer's disease and vascular dementia of Binswanger type and multiple small infarct type. *J. Neural Transm.* **105:** 1283-1291.

Topbas, M.T., Brohi, A.R. and Karaman, M.R. (1998). Cevre Kirliligi. T.C. Cevre Bakanligi Yayinlari, Ankara.

Turkdogan, M.K., Testereci, H., Akman, N., Kahraman, T., Kara, K., Tuncer, I. and Uygan, I. (2003). Dietary nitrate and nitrite levels in an endemic upper gastrointestinal (esophageal and gastric) cancer region of Turkey. *Turk. J. Gastroenterol.* **14:** 50-53.

Tuzel, Y., Gul, A. and Dasgan, H.Y. et al. (2005). Ortu Alti Yetistiriciliginde Gelismeler. In: Turkiye Ziraat Mühendisligi VI. Teknik Kongresi, 3-7 Ocak, Ankara, pp. 551-563.

Tzung, W-J., Po, W-Y. and Wu, J.T. et al. (1995). Effects of some environmental factors on nitrate content of Chinese cabbage *(Brassica chinensis L.). J. Chin. Agric. Chem. Soc.* **33:** 125-133.

Ustyugova, I.V., Zeman, C., Dhanwada, K. and Beltz, L.A. (2002). Nitrates/nitrite alter human lymphocyte proliferation and cytokine production. *Arch Environ. Contam. Toxicol.* **43:** 270-276.

Van Diest, A. (1986). Means of preventing nitrate accumulation in vegetables and pasture plants. In: Fundamentals, Ecological and Agricultural Aspects of Nitrogen Metabolism in Higher Plants. Neeteson, L.H. and Stalen, J.J. (Eds.). Durdrecht: Martins Nijhoft.

Vieira, I.S., Vasconselos, E.P. and Monteiro, A.A. (1998). Nitrate accumulation, yield and leaf quality of turnip greens in response to nitrogen fertilisation. *Nutr. Cycl. Agroecosys.* **51:** 249-258.

Weiss, W.P. (2004). Nitrate toxicity in drought-stressed forages.

www.wvu.edu/agexten/forglvst/nitrate.htm Extension Service West Virginia University.

Whitehead, D.C. (1995). Grassland nitrogen. CAB International, Wallingford, Oxon, UK.

Xu, C., Wu, L.H. and Ju, X.T. et al. (2005). Role of nitrification inhibitor DMPP (3,4-dimethylpyrazole phosphate) in nitrate accumulation in greengrocery *(Brassica campestris* L. ssp. *chinensis)* and vegetable soil. *J. Environ. Sci.* (China) **17:** 81-83.

Yilmaz, I., Ozkan, B., Akkaya, F., Yilmaz, S. and Kutlar, I. (2000). Antalya Ili Sera Sebzeciliginde Ilac ve Gubre Kullaniminin Analizi. *In:* Turkiye 4. Ulusal Tarim Ekonomisi Kongresi, 6-8 Eylul, Tekirdagi, pp. 1-10.

Yilmaz, H. (2004). Turkiye'de Kimyasal Gubre Uretim, Tuketim ve Dis Ticaretindeki Gelismeler. *In:* Turkiye 3. Ulusal Gubre Kongresi, Tarim-Sanayi-Cevre, 11-13 Ekim, Tokat, pp. 35-46.

Yilmaz, H. (2005) Kimyasal Gubre Kullaniminin Cevresel Etkilerinin Teknik ve Ekonomik Acidan Degerlendirilmesi. *In:* I. Cevre ve Ormancilik Surasi, Tebligler, 21-24 Mart, Antalya, pp. 118-128.

Zhou, Z.Y., Wang, W.J. and Wang, J.S. (2000). Nitrate and nitrite contamination in vegetables in China. *Food Rev. Int.* **16:** 61-76.

Zialdi, N., Simard, R.R., Allard, G. and Parent, G. (2000). Yield response of forage grasses to N fertilizer as related to spring soil nitrate sorbed on anionic exchange membranes. *Canad. J. Soil Sci.* **80:** 203-212.

3

Pathophysiology of Nitrate Toxicity in Humans and its Mitigation Measures

Sunil Gupta[1], R.C. Gupta[2] and A.B. Gupta[3]

[1]*Consultant Pediatrician & Neonatologist and Scientist of Environmental Medicine Krishna Ram Ayurvigyan Shodh Sansthan, Jaipur – 302 015, India*

[2]*Professor and Head, Department of Physiology, SMS Medical College, Jaipur – 302 004, India*

[3]*Professor of Environmental Engineering, MNIT, Jaipur – 302 017, India*

e-mail: drsunil@gmail.com

ABSTRACT

Naturally occurring nitrate levels in surface and groundwater are generally a few milligrams per liter. However, nitrate concentrations are higher in watersheds that have been intensively used and modified by human activity. Even in many groundwater's, higher levels are found due to sewage percolating through sandy strata and also due to an excessive use of chemical fertilizers. Excessive intake of drinking water with nitrate content beyond permissible limit (100 ppm) may cause clinical manifestations of disease known as methemoglobinemia (blue baby syndrome) in infants. In adults intake of high nitrate water may cause stomach disorder or gastrointestinal disease. Recent research by the authors has yielded results to explain the mechanism of nitrate toxicity, the defense system in human body to counteract this toxicity and some other manifestations of nitrates apart from causing methemoglobinemia (recurrent stomatitis, recurrent diarrhea and Recurrent Respiratory Tract Infection) among children, which could be of high importance to the field of environmental health. The current chapter critically discusses the pathophysiology of nitrate toxicity in humans and suggests potential measures for its mitigation in detail.

Keywords: Nitrate, Toxicity, Human health, Mitigation

INTRODUCTION

Naturally occurring nitrate levels in surface and groundwater are generally a few milligrams per liter. However, nitrate concentrations are higher in watersheds that have been intensively used and modified by human activity. Even, in many groundwaters, higher levels are found due to sewage percolating through sandy strata and also due to an excessive use of chemical fertilizers. World Health Organization (WHO) report, 2004, maintains that extensive epidemiological data support limiting the value of nitrate-nitrogen to 10 mg/L or as nitrate to 50 mg/L (WHO 2004) for human consumption whereas IS-10500 prescribes maximum permissible limits of nitrate in drinking water as 45 mg of NO_3 per liter (IS 10500 1995).

The problem of nitrates is endemic internationally as well as nationally, but so far no compiled data at national level are available. The data available are either of small areas or of scattered zones/areas. Many workers tried to compile the status of nitrate in drinking water in Rajasthan, India (PHED Habitat Survey 1991-93; Kumar et al. 2002).

EPIDEMIOLOGY

The problem of nitrate toxicity is global and has been reported from a number of countries (Prakasa Rao and Puttanna 2006).

International Perspective

Nitrate is a widespread contaminant of ground and surface waters worldwide (Prakasa Rao and Puttanna 2006; Singh et al. 1994). In the United States, the problem appears to be concentrated in the Mid-West and the Far-West, with large areas of Iowa, Illinois, Kansas, Michigan, Wisconsin, Washington and California being heavily affected (Smith et al. 1994; Revenga and Mock 2001). The USGS (Smith et al. 1994) reported that nitrate concentration in the nation's groundwater supply was increasing steadily. In South America, nitrate concentrations in the monitored watersheds are relatively low and follow human land use. The highest nitrate concentrations are found in the Uruguay watershed, where intensive agriculture is practiced. Likewise, nitrate concentrations in water are also greater in the Magdalena watershed of Colombia than in the less densely populated watersheds of the Amazon basin. The nitrate concentrations in South America correspond to lower fertilizer application rates in comparison to Europe.

Nitrate concentrations are higher in watersheds that have been intensively used and modified by human activity such as the Weser, Seine, Rhine, Elbe, and Senegal. High levels are also found in such watersheds in China, South Africa, and the Nile and Mississippi basins. South Africa has some of the highest natural nitrate levels in the world (more than 550 mg/L NO_3^--N) especially in the Kalahari region (Colvin et al. 2008).

National Perspective

Nitrate from sources like mineralized organic matter, fertilizers, organic manures, industrial effluents or urban wastes results in groundwater pollution, which has received attention in various parts of India. The available data from various studies point to the fact that nitrate pollution from any of the above sources can pose a serious health hazard to humans, animals, birds and aquatic life. The data available on the concentrations of nitrate in groundwater samples from Punjab, Haryana, U.P., Delhi, Orissa, Tamil Nadu, Bihar, A.P., M.P., Maharashtra, Karnataka and other states point to the fact that in many of these samples, the nitrate content in groundwater has been much more than the permissible limits. In many parts of Punjab and Haryana, nitrate level in the groundwater is increasing continuously beyond the critical standard. Nitrate levels in the groundwater over vast agricultural areas can be correlated with intensive irrigated agriculture, use of nitrogenous fertilizers, and groundwater exploitation. Workers tried to evaluate the nitrate pollution and its strategies for reducing it (Prakasa Rao and Puttanna 2006). Nitrate pollution of groundwater by fertilizer use in India has also been reviewed (Singh et al. 2007).

SOURCES

There are two sources of nitrogen nitrate for human beings: exogenous sources and endogenous sources.

Exogenous Sources

Wastes containing organic nitrogen are decomposed in soil or water by microbial action forming ammonia, which is then oxidized to nitrite and nitrate. Since nitrite is easily oxidized to nitrate, it is primarily nitrate that is found in groundwater and surface waters. Nitrate-containing compounds in the soil are generally soluble and readily migrate with the groundwater. Nitrosamines have a short lifespan in ambient air. Vegetables (cauliflower, spinach, collard greens, broccoli, and root vegetables) account for more than 70 percent of the nitrates ingested in the human diet. The remainder of nitrate in a typical diet comes from drinking water (21 percent) and meat and meat products (6 percent). Nitrates are naturally occurring inorganic ions, which are part of the nitrogen cycle, and are commonly found in groundwater and surface waters. The nitrate ion (NO_3^-) is the most stable form of nitrogen in oxygenated environments, thus all nitrogen-containing molecules can act as sources of nitrates.

Water and food

Nitrogen is an essential major element for plant growth (Allan Wild, 1988). Nitrogen nutrition of crops is very often the limiting factor for food production and nutritional quality of food as indicated by its protein content. Thus, nitrogen has a key role to play in feeding the world population. Since a large response to nitrogenous fertilizers is a universal phenomenon, farmers usually apply large doses of fertilizer. But the recovery of applied nitrogen by the crop rarely exceeds 50% and a large fraction of the surplus nitrogen gets transformed into nitrate form and pollutes the environment including surface and groundwaters. The interaction between the various forms of nitrogen in soil, plants, animals and atmosphere constitutes the nitrogen cycle (Fig. 3.1).

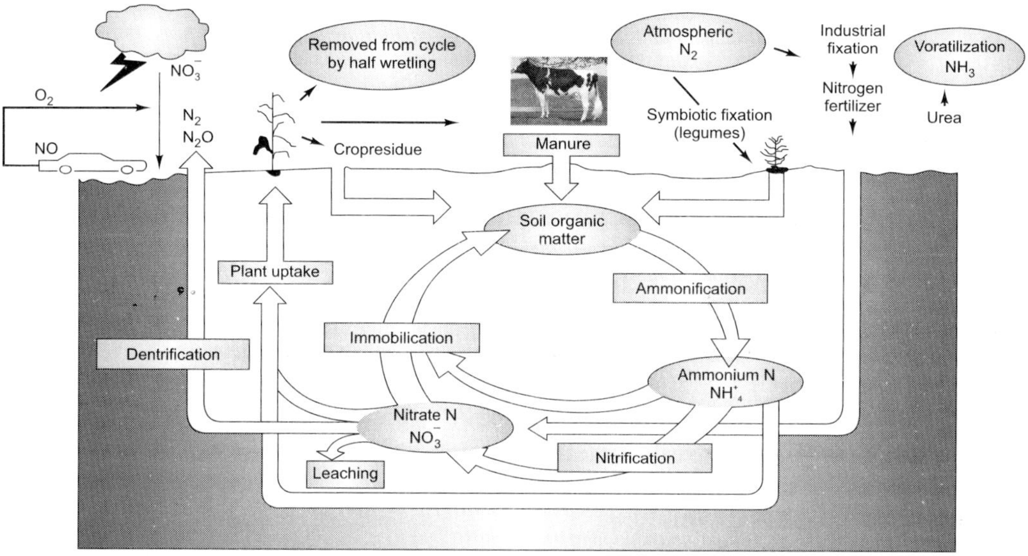

Fig. 3.1. Nitrogen in agricultural system (Wolkoswki et al., Internet publication).

Leafy vegetables

Being a rich source of nutrients and antioxidants, leafy vegetables occupy an important place in the human diet. Some vegetables are known to accumulate high concentration of nitrate under the current practice of heavy fertilization. Vegetables are the major source of daily intake of nitrate by human beings, supplying about 72 to 94% of the total intake (Annexure I). On consumption by humans, part of this nitrate is converted to nitrite and N-nitroso compounds that have detrimental effects on human health.

The nitrate content in the vegetables available in the markets is reaching as high as 4451 mg kg^{-1} fresh weight in chenopodium and 4293 mg kg^{-1} fresh weight in spinach (Anjana et al. 2007a,b).

Other sources of nitrate pollution

Malik (2000) suggested that the presence of nitrate bearing rocks in geological formations as in some parts of Haryana contributed to high nitrate levels in groundwater (Malik 2000) (Table 3.1).

Table 3.1: Other sources of nitrate pollution (Malik RPS 2000)

	Agriculture	Municipal	Industrial
Diffuse sources	Use of synthetic nitrogen fertilizers	Combustion engines in vehicles	Atmospheric emission (oxides of nitrogen) from combustion engines and energy production
	Use of organic manures	Disposal of municipal effluents	Disposal of effluents
Point sources	Accidental spills	Badly designed landfills	
	Improper, shelter less composting pits, heaps	Septic tanks	Disposal of wastes into wells, ponds, injection into tube wells, etc.
		Leaking sewerage systems	Badly designed landfills
Linear sources	Contamination of rivers with major groundwater connections		

Contamination from agro-based industries

The methods of disposal of industrial effluents include diversion to surface water sources (rivers, lakes, canals, etc.), stored in lagoons or ponds for percolation, or spreading or spraying onto the ground surface. The effluents may be treated, partly treated or untreated. The polluted water percolates either directly from the surface water bodies or agricultural fields (when such water is used for irrigation), and pollutes the groundwater. The extent of pollution depends upon the volume of effluents discharged per day and their characteristics (Wakida et al. 2004). Under certain circumstances, forests also contribute to nitrate pollution.

Continuous application of nitrogen fertilizers can cause an increase in the concentration of nitrates in fruits, vegetables and food. Maximum admissible level of nitrites in vegetables is 1 mg/kg (Mondal et al. 2008).

Cooking in aluminum utensils increases the reduction of nitrates to nitrite (WHO 1985a), hence increasing the intake.

Endogenous Sources

Nitrates, known as endogenous nitrate, are also produced in the body. A major pathway for endogenous nitrate production is conversion of arginine by macrophages to nitric oxide and citrulline, followed by the oxidation of nitric oxide to nitrous anhydride and then reaction of nitrous anhydride with water to yield nitrite. Gastrointestinal infections and non-specific diarrhea increase endogenous (non-bacterial) nitrate synthesis, probably induced by the activation of mammalian reticuloendothelial system (WHO 1985b; WHO 1996).

Total Daily Intake of Nitrate

Daily dietary intakes of nitrate and nitrite have been estimated in different countries (Ellen and Schuller 1983). The variation in the quantity of nitrates and nitrites ingested via diet is extremely high. For example, individuals who seldom eat vegetables and cured meats have a low intake, whereas vegetarians have a relatively high intake. In most European countries, the mean nitrate intake is about 10–30 mg/day. Vegetarians usually have a two-to fourfold higher intake of nitrates than non-vegetarians (WHO 1985c).

KINETICS AND METABOLISMS

About 20% of the ingested nitrate is reduced to nitrite by nitrate-reducing microflora present in the saliva (Eisenbrand et al. 1980) at the base of the tongue (Walker 1995). The factors which influence the oral microflora and hence the reduction of nitrate are nutritional status, infection, environmental temperature and age (more in elderly) (Eisenbrand et al. 1980).

Ingested nitrate is reduced to nitrite by nitrate reducing microflora in the stomach (in favorable conditions, viz. pH ≥ 4) and upper part of intestine and may be at other parts of the human gastrointestinal tract. In normal conditions, the reduction of nitrate to nitrite does not occur in the stomach, but in situations where the stomach pH is high, such as achlorhydria (Ruddell et al. 1978), atrophic gastritis (Walker 1995; Mirvish 1975), artificially fed infants, or patients using antacid or similar drugs, e.g. omeprazole (Farinati et al. 1996; Colbers et al. 1996; Vermeer et al. 2001), the conversion of nitrate to nitrite occurs even in stomach. The high pH of the stomach favors the growth of nitrate reducing organisms.

This nitrite is readily and completely absorbed from both the stomach and the upper small intestine. Approximately 25% of the ingested nitrate is actively secreted into saliva, where it is partly (20%) reduced to nitrite by the oral microflora; nitrate and nitrite are then swallowed and reenter the stomach.

Absorbed nitrite is then rapidly distributed throughout the tissues. Absorbed nitrite is rapidly oxidized to nitrate in the blood, with the formation of methemoglobin.

The following is the flow chart of the fate of ingested nitrate (Li et al. 1997; WHO 1977a) (Fig. 3.2).

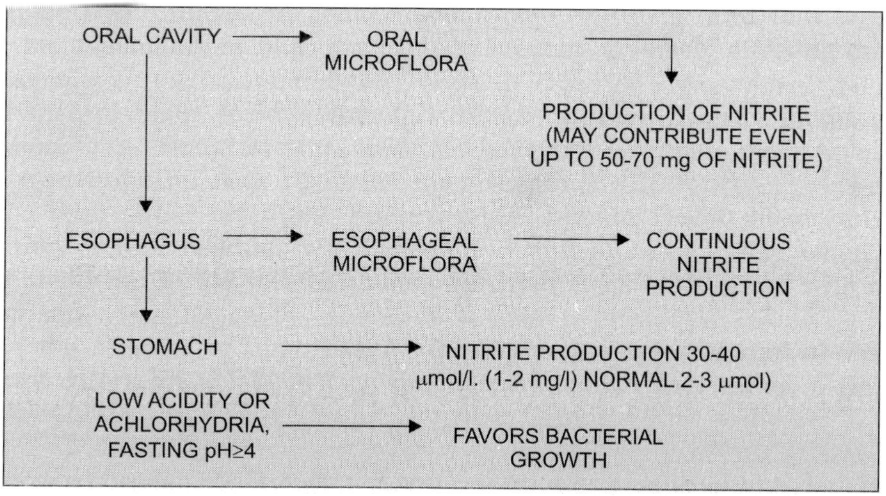

Fig. 3.2. Fate of ingested nitrate (Li et al., 1997; WHO 1977a).

Formation of N-Nitroso Compounds

Formation of N-nitroso compounds is a multiple step process (Choi 1985). First, nitrate is converted into nitrite after consumption. Then the nitrite reacts with natural or synthetic organic compounds (known as secondary amines or amides) in food or water to form new combinations called N-nitroso compounds (either nitrosamines or nitrosamides). Many of these N-nitroso compounds have been found to be carcinogenic in all the animal species tested, although some of the most readily formed compounds, such as N-nitrosoproline, are not carcinogenic in humans. At least 75% of the 120 N-nitroso compounds have been found to be carcinogenic to animals (Gilli et al. 1984; Terblanche 1991). The most common N-nitroso compounds are dimethylnitrosamine (DMN), N-methylmethanamine (DMA), trimethylamine (TMA) and trimethylamine oxide (TMAO). The N-nitroso compounds are carcinogenic in animal species and are probably also carcinogenic to humans. The data from a number of epidemiological studies are at most, only suggestive, relating to carcinogenicity in humans, but it has been reported that a link between cancer risk and endogenous nitrosation as a result of high intake of nitrate and, or nitrite and nitrosatable compounds is possible (RIVM 1989; WHO 1996).

Predisposing Factors

A direct correlation between gastric pH, bacterial colonization and gastric nitrite concentration has been observed in healthy people (Mueller et al. 1986). In individuals with gastrointestinal disorders and achlorhydria, high levels of nitrite have been reported (Ruddell et al. 1978; Dolby et al. 1984). Infections and non-specific diarrhea played a role in the increased endogenous synthesis of nitrate (Lee et al. 1970; Tannenbaum et al. 1978; Green et al. 1981; Hegesh and Shiloah 1982; Bartholomew and Hill 1984; Gangolli et al. 1994).

Trans-Placental Cross

Nitrite has been shown to cross the placenta and cause the formation of fetal methaemoglobinemia in rats (El Nahas et al. 1984).

Half-Life

The half-life of nitrate in the body after ingestion is approximately 5 h (Wagner et al. 1983a). Nitrite was not detected in any of the body fluids studied except saliva where it appeared to increase as the nitrate levels decreased (Cortas and Wakid 1991).

METABOLISM OF INGESTED NITRATE IN HUMAN BODY AT CELLULAR LEVEL

Ingested inorganic or organic nitrates will result in increased oxidation of hemoglobin to methemoglobin and increased production of nitric oxide (Murray et al, 1993;

Waldman et al, 1987; Craven et al. 1978) (Figure 3.1). The conversion of nitrite to nitric oxide is a non-enzymatic process. (Vane et al. 1994; Lowestein et al. 1994; Robertson 1996; Smith et al. 1997). The oxidation of hemoglobin to methemoglobin results in the formation of superoxide radical by the transfer of a single electron. The enzyme, superoxide dismutase, present in the erythrocytes, catalyses the conversion of superoxide radical ($O^{\bullet -}$) to H_2O_2 and O_2. The H_2O_2 is decomposed by glutathion peroxidase or catalase both are present in erythrocytes (Sutton et al. 1976; Winterbourn et al. 1976). Once the rate of oxidation of hemoglobin increases sufficiently in erythrocytes and overwhelms the protective and reductive capacities (e.g. cytochrome b_5 reductase system etc.) of the cells (Bodansky 1951; Jaffe 1981), there is an increased production of reactive free radicals of nitric oxide ($NO^{\cdot}$) and oxygen (Winterbourn et al. 1976).

Fate of Free Radical Nitric Oxide ($NO^{\cdot}$)

Hemoglobin scavenges nitric oxide through the high affinity ferrous sites on heme to form S-nitrosothiol, which has an affinity to nitric oxide 8000 times higher than that for oxygen (Hsia 1998) by binding at β 93 cysteine residue on the globin chain. As hemoglobin binds oxygen in the lungs, its binding affinity to S-nitrosothiol is increased. As hemoglobin releases oxygen at the periphery, its affinity for S-nitrosothiol is reduced and nitric oxide is released in the tissues (Hsia 1998). The thiol group of S-nitrosothiol protects nitric oxide from being scavenged by the binding site on heme. Thus, in addition to carrying oxygen, hemoglobin acts as a carrier of nitric oxide. The enhanced release of nitric oxide from S-nitrosohemoglobin in hypoxic tissue in turn reduces regional vascular resistance.

Nitric oxide is a biogenic messenger, an endothelial derived relaxing factor (EDRF) (Jaffe 1981; Hsia 1998)) and activates guanylyl cyclase system (Berger et al, 1997) [converts guanosine triphosphate (GTP) to 3′5′ cyclic guanosine monophosphate (cGMP)], raising the cGMP pool and therefore inducing *inter alia* vasodilatation (Berger et al. 1997) by lowering intracellular calcium ion (Smith et al. 1997).

Fate of Free Oxide Radical ($O^{\cdot}$)

In a normal cell, $O^{\bullet -}$ is scavenged by the enzyme superoxide dismutase, and H_2O_2, which is a product of reaction is removed by glutathion peroxidase and catalase (Roediger et al. 1986; Comly 1945). Any $O^{\bullet -}$ that escapes this mechanism leads to the production of much more reactive substances such as hydroxyl radical ($OH^{\bullet}$) and peroxynitrite radical. $O^{\bullet -}$ and H_2O_2 are highly selective in their reaction with biological molecules, whereas $OH^{\bullet}$ attaches everything around it and reacts with other cell constituents, causing irreversible cell damage. This mechanism is likely to become more significant if $O^{\bullet -}$ is produced in abnormally high amount (e.g. excessive nitrate ingestion), or if any of the protective mechanisms are defective (Sutton et al. 1976; Hsia 1998).

Thus, increased consumption of nitrate will lead to (a) increased production of nitrite (Allison et al. 1984; Cole et al. 1980); (b) enhanced absorption of sodium from the intestinal lumen (Roediger et al. 1986); (c) excess $NO^{\bullet}$ (free radical nitric oxide) generation, having vasodilatory effect (Winterbourn et al. 1976; Hsia 1998; Berger et al. 1997; Nitric Oxide 1997), and (d) increased production of $O^{\bullet -}$, which will react with other cell constituents, possibly causing irreversible cell damage (Berger et al. 1997; Gupta et al. 1998) (Fig. 3.3).

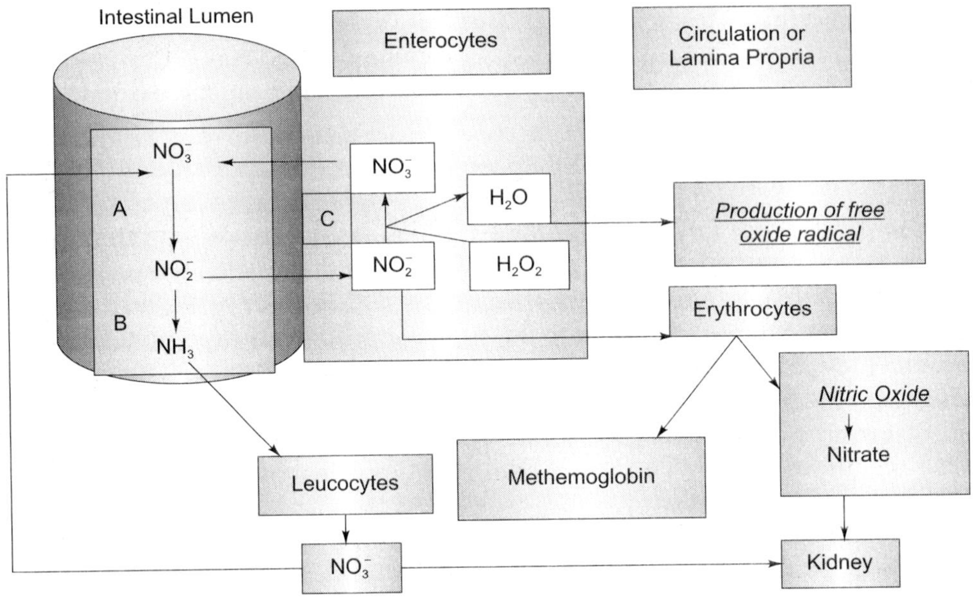

Fig. 3.3. Metabolism of ingested nitrate in human body at cellular level. A – Bacterial nitrate reductase; B – Bacterial nitrite reductase; C – Catalase

EXCRETION

About 70–75% of ingested nitrate is excreted within 24 hours of ingestion irrespective of the amount of intake. The excretion was noticed high in the first 5 hours. (Bartholomew and Hill, 1984; Wagner et al. 1983b). However, various authors have reported that urinary nitrate excretion may exceed nitrate intake even in infants if the latter is low, as a consequence of endogenous nitrate formation (Hegesh and Shiloas, 1982). Urinary nitrate excretion in infants was reported to be 80–100% of the average intake, but no specific data were given for exposure levels (Turek et al. 1980).

Low levels of nitrate and nitrite were detected in the feces of humans (Saul et al. 1981).

A small amount of ingested nitrate converting to nitrite, changes to oxides of nitrogen (nitric oxide, nitrous oxide, nitrogen, etc.) in blood and gets excreted by exhalation through lungs (as shown in the flow chart given below).

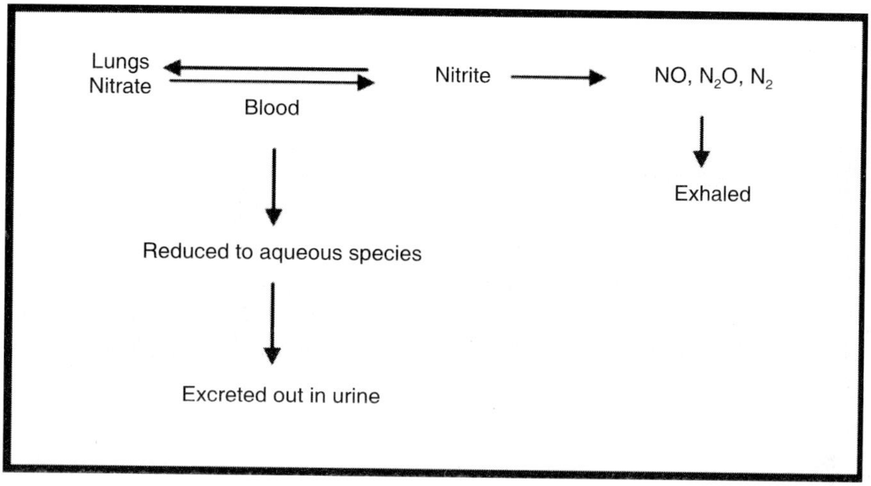

ACUTE TOXIC EFFECTS

Exposure to high nitrate may cause acute toxicity due to acute exposure.

In Humans

Human lethal doses of 4–50 g NO_3^- (equivalent to 67-833 mgNO_3^-/kg BW) have been reported. Toxic doses with methemoglobin formation as a criterion for toxicity ranged from 2 to 5 g (Corré and Breimer, 1979) of NO_3. These values are equivalent to 33–83 and 100–150 mg NO_3^-/kg body weight respectively.

Signs and symptoms of acute nitrate toxicity: Fassett (1973) reported a rapidly occurring severe gastroenteritis with abdominal pain, blood in the urine and feces as symptoms of acute nitrate intoxication. Repeated doses gave rise to dyspepsia, mental depression, headache and weakness. Cyanosis (bluish discoloration) is also present.

In Animal Studies

The studies relating to acute exposure of nitrate were conducted in animals. The acute oral toxicity of nitrate to laboratory animals is low to moderate. LD50 values of 1600–9000 mg of sodium nitrate per kg of body weight have been reported in mice, rats, and rabbits (Til et al. 1988). Ruminants are more sensitive to the effects of nitrate as a result of high nitrate reduction in the rumen; the LD50 for cows was 450 mg of sodium nitrate per kg of body weight. Nitrite is more toxic than nitrate: LD50 values of 85–220 mg of sodium nitrite per kg of body weight have been reported for mice and rats (RIVM, 1989; WHO, 1996).

CHRONIC TOXIC EFFECTS

It is caused due to long-term exposure of non-lethal doses. The following effects have been reported:

Methemoglobinemia

Nitrates in drinking water have been reported to cause methaemoglobinemia in infants up to 6 months of age.

Mechanism of methemoglobin formation

The essential action in the formation of methemoglobin is an oxidation of ferrous to ferric ion. This oxidation may be brought about in one of the following ways (Bodansky, 1951): by the direct action of the oxidants, or by the action of hydrogen donors in the presence of oxygen, or by auto oxidation. In the presence of nitrites, the ferrous ion of hemoglobin gets directly oxidized to ferric state. Normally, the methemoglobin that is formed is reduced to hemoglobin by cytochrome b_5/ cytochrome b_5 reductase system as follows (Benz, 2001):

$$Hb^{+3} + \text{Red cyt } b_5 \longrightarrow Hb^{+2} + \text{oxy cyt } b_5$$

Oxidised cytochrome b_5 (Red Cyt b_5) is regenerated by the enzyme cytochrome b_5 reductase:

$$\text{oxy cyt } b_5 + \text{NADH} \xrightarrow[\text{Reductase}]{\text{cyt } b_5} \text{Red cyt } b_5 + \text{NAD}$$

Cytochrome b_5 reductase is a NADH dependent enzyme. NADH is supplied from glycolysis (Fig. 3.4).

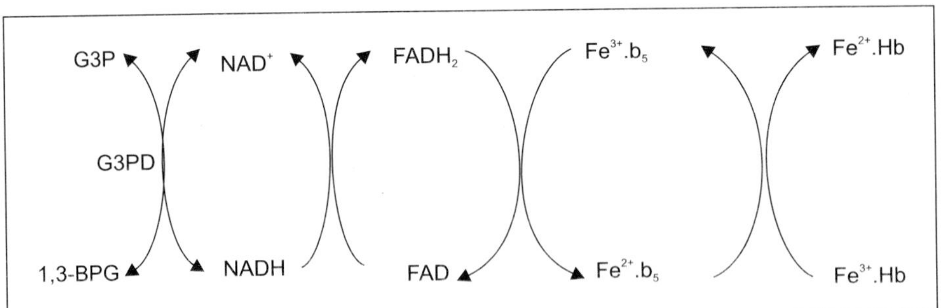

Fig. 3.4. Reduction of methemoglobin (Nagel 2007).

Thus, the enzyme cytochrome b_5 reductase plays a vital role in counteracting the oxidative effects of nitrate ingestion. However, permissible concentration (50 mg/L as nitrate as per (WHO, 2004) normally present in the water or food does not cause any health risk to adults, but infants constitute a vulnerable group for the following reasons (WHO 1977b):

- Relatively higher stomach pH (2.0–5.0), which permits the growth of nitrate reducing organisms such as *Coliforms, E. coli, Pseudomonas fluorescence, B. subtilis, Staph. albus*
- Relatively higher consumption of water per unit weight of body

- The presence of fetal Hb, which readily gets oxidized to methHb.
- Poorly developed cytochrome b_5 reductase system
- Nitrate gets concentrated by repeated boiling of water for feeding
- Bacterial contamination of the water itself or dried milk powder
- Early weaning on to nitrate rich vegetables, e.g. spinach
- Diarrhea causes increase in stomach pH

In India, breastfeeding, which is a common practice especially up to the age of 6 months, protects the infants from nitrate toxicity. The probability of toxicity to fetus in pregnant women cannot be ruled out as transplacental passages of nitrate metabolite (nitrite) have been documented in animal experiments.

While a few cases of methemoglobinemia in infants have been reported to be associated with water nitrate levels of less than 50 mg nitrate ion/L, most cases occur with nitrate level of 90 mg nitrate ion/L or more (Comly 1945; Cornblath et al. 1948; Jaffe 1981; Marshall et al. 1945; Knotek et al. 1964; WHO 1977a). Recently methemoglobinemia has been reported in all age groups (Gupta et al. 2000). It was observed that methemoglobinemia was prevalent among all age groups, but was more severe in infants and older age groups (> 45 years).

Once methemoglobin levels in the blood exceeds 10% of total hemoglobin, it manifests as clinical cyanosis and causes cellular anoxia. The effects of methemoglobin (in relation to %) and clinical presentation are given below (WHO 1977c):

Meth-Hb	Clinical presentation
<10%	No signs and symptoms
10-25%	No symptoms, cyanosis present
25-50%	Cyanosis, dyspnoea, headache
50-60%	Dyspnoea even on lying, cyanosis, disorientation
60%	Lethal levels

Cytochrome b_5 Reductase Adaptation

In several Indian villages, people have been consuming water containing high nitrate concentrations, at times up to 1200 mg nitrate ion/L. High nitrate concentrations are causing methemoglobinemia in all ages (Gupta et al. 2000). It was observed that this methemoglobinemia was more in infants and elderly people (Gupta et al. 2000). Children and adolescents have lower levels of methemoglobin. Adaptation to the reserve of cytochrome b_5 reductase activity with increasing water nitrate concentration to compensate methemoglobinemia was reported to be a possible mechanism. The study (Gupta et al. 1999b) indicated that high nitrate concentrations are hazardous not only to infants but also to groups over 18 years of age. It was reported that adaptation of cytochrome b_5 reductase activity peaks at about 95 mg nitrate ion/L nitrate concentration and gets exhausted by nitrate level 200 mg nitrate ion/L, hence making people more prone to toxicity.

Infant Mortality Rate (IMR)

A study (Super et al. 1981) on African mothers and other studies also (Spalding and Exner 1993; CDC 1996; Dorsch et al. 1984; Fewtrell 2004; Schwiede 2005) reported an increase in infant deaths with increasing exposure of pregnant mothers and infants to nitrate. This may either be due to undetected toxic methemoglobinemia or malformations and weaknesses in the infant caused by fetal nitrate exposure.

It was further suggested that because of high IMR there is a need to revise the nitrate standards for drinking water (Kumar S. et al. 2002).

Nitrate, Nitrite, Nitrosamines and Cancer

Nitrate itself is not carcinogenic, but instead acts as a "procarcinogen", meaning that it reacts with other chemicals (amines and amides) to form carcinogenic compounds (N-nitroso compounds). Under acidic conditions, nitrites (NO^{2-}) are formed naturally from nitrates, and nitrites, in turn, may combine with amines or amides to form N-nitroso compounds (nitrosamines). The endogenous formation of carcinogenic N-nitroso compounds can occur following ingestion of nitrate from drinking water. Nitrate is first reduced in the saliva to nitrite, which can react in the stomach with secondary amines and amides. The nitrosamines most commonly reported in foods are dimethylnitrosamine, diethylnitrosamine, nitrosoproline and nitrosopyrrolidine. Nitrosamines can also be found in some alcoholic beverages and tobacco products.

In animal or human studies, N-nitroso compounds have been associated with 15 different types of cancers, including tumors in the bladder, stomach, brain, esophagus, bone and skin, kidney, liver, lung, oral and nasal cavities, pancreas, peripheral nervous system, thyroid, trachea, acute myelocytic leukemia, and T and B cell lymphoma. These group of N-Nitroso compounds are found to be carcinogenic in a wider range of tumors than any other group of carcinogens (NAS 1977, 1978; IARC 1978; Mirvish, 1991 and 1983). More than one hundred of these N-nitroso compounds have been tested for carcinogenicity in animals, and 75–80% of them have been found to be carcinogens (NAS 1977). IARC (1978) concluded that 11 N-nitroso compounds were carcinogenic in man. In humans, the organs thought to be most at risk from cancer are the stomach, esophagus, nasopharynx, and bladder.

To summarize the carcinogenesis due to high nitrate ingestion is multifactorial. The possible mechanisms may be:

1. Increased formation of nitrosamines (known carcinogen), a metabolic intermediate product of nitrate metabolism.
2. Increased formation of free oxide radicals
3. Interference in NO metabolism

Still there are studies (Hill et al. 1973; Armijo et al. 1975; Cuello, 1976; Zaldivar et al. 1977; RCEP 1979; Fraser et al. 1980) which have drawn no firm conclusion to prove nitrate as a carcinogen.

Common cancers reported due to increased consumption of nitrates are:

1. Gastric cancer (Jaskiewicz et al. 1990; Farinati et al. 1989; Vermeer et al. 2001; Dallinga, 1998)
2. Non-Hodgkins lymphoma (NHL) (Weisenburger, 1990)
3. Colorectal cancer (De Roos et al. 2003 ; Vander et al. 1994)
4. Urinary bladder cancer (Michaud, 2001; Weyer et al. 2001).

Pathophysiology of carcinogenesis due to NO and free oxide radicals

Cancer is the endpoint of a multistep process that includes three fundamental components: initiation, promotion, and progression (Fig. 3.5). Arginine is involved in a number of biosynthetic pathways that significantly influence carcinogenesis and tumor biology. Since the discovery that arginine metabolism generates a ubiquitous signal transduction molecule, nitric oxide (NO), 3 arginine-derived NO has been found to play a significant role in many of the specific events that lead to cancer.

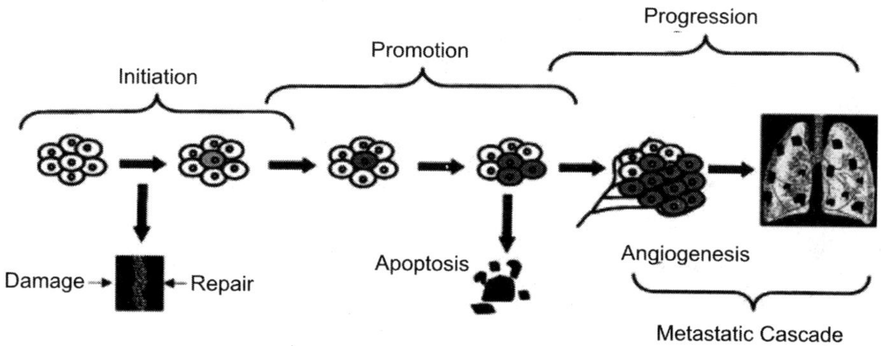

Fig. 3.5. Simplified diagram of the multi-step process resulting in cancer.

Hence, there is a possibility of cancer development in the conditions where NO production is enhanced, e.g.,

1. High nitrate ingestion as has been observed, increases the rate of colorectal cancer
2. Use of silenafil citrate
3. Drugs which have good amount of L arginine
4. Inhibition of cancer growth by NOS inhibitor L-NAME.

Respiratory System

A correlation between drinking water nitrate concentration and pathological changes in bronchi and lung parenchyma has been reported in animal studies (Shuval et al. 1972; Gruener et al. 1970). Changes in the lungs reported were frequent dilation of bronchi with lymphocytic infiltration of mucosa and muscles, frequent purulent bronchial exudates, interstitial round cell infiltration and fibrosis at certain areas.

WHO (1977d) reported an association of increased asthmatic attacks and high air-borne nitrate concentrations. A high percentage (40–82%) of cases of acute respiratory tract infection with history of recurrence has been reported in children drinking high nitrate in water (Gupta et al. 2000). These findings were further substantiated (Gupta et al. 1999c) in an animal experiment on rabbits.

NOx, tobacco and malignancy

Tobacco use causes 20% of cancer deaths worldwide. The International Agency for Research on Cancer predicts 10 million tobacco-related deaths annually by 2020, of which 70% will occur in the developing world (IARC, 2004).

This study conducted by Sleiman et al. (2010) reported that residual nicotine from tobacco smoke sorbed to indoor surfaces reacts with ambient nitrous acid (HONO) to form carcinogenic tobacco-specific nitrosamines (TSNAs). Substantial levels of TSNAs were measured on surfaces inside a smoker's vehicle. Time-course measurements revealed fast TSNA formation, with up to 0.4% conversion of nicotine. Given the rapid sorption and persistence of high levels of nicotine on indoor surfaces—including clothing and human skin—this recently identified process represents an unappreciated health hazard through dermal exposure, dust inhalation, and ingestion. Whereas direct inhalation of SHS is an exposure pathway of concern, nonsmokers, especially infants, are at risk through contact with surfaces and dust contaminated with residual smoke gases and particles (Matt et al. 2004).

N-nitrosamines were found to be unstable in sunlight, rendering the reaction unimportant in outdoor daytime conditions. However, this process can be relevant indoors where N-nitrosamines and HONO are less vulnerable to photochemical decomposition.

Nitrate and animal experiments

Rabbit studies in lungs (Gupta et al. 1999) indicated no damage at 100 ppm nitrate in drinking water. As the nitrate concentration in water increases to 400 ppm (Fig. 3.6), the changes in lungs indicated in the form of congestion, presence of inflammatory cells and breakdown of alveoli.

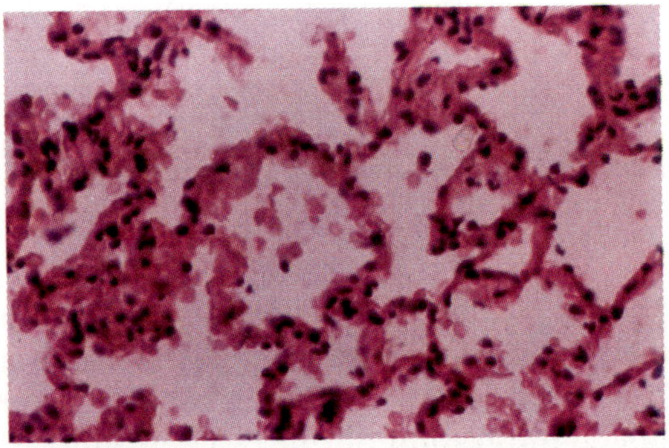

Fig. 3.6. Histopathological changes in lung at 400 ppm (Hematoxylin and eosin, x 400, Gupta et al. 1999).

CARDIOVASCULAR SYSTEM

The cardiac toxicity of nitrate is mediated through its metabolic effects and metabolic products, e.g., methemoglobin formation, increased NO formation, increase in free oxide radicles (ROS) and high levels of compensatory enzymes (Cytochrome b_5 reductase etc.). The following effects have been reported:

1. Early onset of hypertension
2. Inflammation and degeneration of cardiac musculature
3. Causing endothelial dysfunction
4. Making the person prone for ischemia
5. Enhances arteriosclerotic and atherosclerotic process

Early Onset of Hypertension

Earlier onset of hypertension has been reported with high nitrate ingestion (Malberg, 1978). Pomeranz et al. (2000) reported that elevated salt and nitrate levels in drinking water cause an increase of blood pressure in school going children.

Inflammation and Degeneration of Cardiac Musculature

The nitrate toxicity on cardiovascular system has been reported in both animals and human beings. Animal studies in early seventies (Shuval et al. 1972; Gruener et al. 1970) reported correlation among drinking water nitrate concentration, high methemoglobin levels and cardiac muscles (Shuval et al. 1972; Gruener et al. 1970). The changes reported in cardiac muscles were small foci of inflammatory cells and fibrosis, diffuse interstitial cellularity with pronounced degenerative foci, intramural coronary arteries resulting in thinning and dilation in comparison to the control group.

In an animal experiment (Gupta et al. 1999), changes in the cardiac muscles with different concentrations of nitrate in water are given below (Table 3.2).

Table 3.2: The effect of nitrate present in the drinking water on cardiac muscles

Water nitrate (mg/L)	Branching of myosites	Inflammatory cells	Focal degenerative changes
45	NIL	NIL	NIL
100	+	NIL	NIL
200	+	NIL	NIL
400	+	+	+
500	+	+	+

(+) Mild, (++) Moderate, (+++) Severe, NIL: Normal histopathology (No change observed)

Figure 3.7 depicts the changes in the cardiac muscle in the rabbits fed with water containing 400 mg of NO_3 ion per liter (Fig. 3.7). The figure shows branching of myosites with focal degenerative changes.

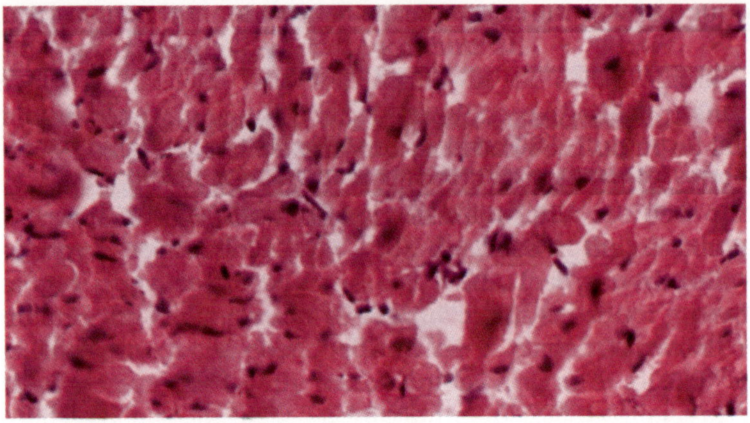

Fig. 3.7. Histopathological changes in the cardiac muscle at 400 ppm (Hematoxylin and eosin, x 400, Gupta et al. 1999).

Endothelial Dysfunction

Endothelial dysfunction, or the loss of proper endothelial function, is a hallmark for vascular diseases, and is often regarded as a key early event in the development of atherosclerosis. An association has been observed between the increase in plasma nitrite/nitrate levels and an excessive production and/or inactivation of NO leading to impaired vascular smooth-muscle reactivity resulting in impairment in endothelium dependent, shear stress-induced vasodilatation.

Making the Person Prone for Ischemia

Nitrate and nitrite have emerged as a viable alternative source of NO under ischemic conditions. NO has been shown to be one of the most important molecules for the prevention of injury from ischemia. Paradoxically, in conditions of inadequate oxygen, the NO is dangerous and may be the cause of ischemia. (Garg and Bryan 2009).

Atherosclerosis

Nitrate is a source of ROS in human body. ROS is known to cause enhanced atherosclerosis. Atherosclerosis develops from low-density lipoprotein molecules (LDL) becoming oxidized (ox-LDL) by free radicals, particularly reactive oxygen species. When oxidized LDL comes in contact with an artery wall, a series of reactions occurs to repair the damage caused to the artery wall by oxidized LDL. The LDL molecule is globular-shaped with a hollow core to carry cholesterol throughout the body. Eventually, the artery becomes inflamed. The cholesterol plaque causes the muscle cells to enlarge and form a hard cover over the affected area. This hard cover causes narrowing of the artery, reduces the blood flow, and increases blood pressure (Duguid 1960).

GASTROINTESTINAL SYSTEM

In gastrointestinal system, four major findings have been reported:

1. Recurrent diarrhea in children
2. Recurrent stomatitis
3. GI malignancy
4. Histopathological changes in animal study

Recurrent Diarrhea in Children

Gupta et al. (2001) reported a problem of recurrent diarrhea in children upto 8 years of age. They suggested that increased consumption of nitrate leads to (a) increased production of nitrite (Allison et al. 1984; Cole et al. 1980); (b) enhanced absorption of sodium from the intestinal lumen (Roediger et al. 1986); (c) excess $NO^{\bullet}$ (free radical nitric oxide) generation, having vasodilatory effect (Winterbourn et al. 1976; Berger et al. 1997; Nitric Oxide 1997; Smith et al. 1997; Hsia 1998); and (d) increased production of O_2^-, which will react with other cell constituents, possibly causing irreversible cell damage (Berger et al. 1997; Gupta 1998). These changes in enteric mucosa cause hyperemia and edema in the enteric mucosa and later on possibly cause irreversible mucosal damage, and therefore provide high-risk conditions suitable for recurrent diarrhea.

These findings are of interest since infants and children, if given nitrate rich water with ORS (Oral Rehydration Solution) during diarrhea, it will act as aggravating factor for nitrate toxicity (Murray et al. 1993), where the use of WHO oral rehydration solution (ORS) is a normal routine. The use of WHO ORS could be of grave concern, if prepared with water containing high nitrate.

Recurrent Stomatitis

Recurrent stomatitis was another problem reported in people using high nitrate containing drinking water. This finding was observed in all groups and well correlated with increased cytochrome b_5 reductase activity following high nitrate ingestion. The increased activity of enzyme cytochrome b_5 reductase is associated with stomatitis (Gupta et al. 1999a). Another possible cause may be the production of NO, which is a metabolic product of nitrates and a known inflammatory agent.

Histopathological Changes in Animal Study

Animal studies conducted by Gupta et al. (2001) indicated changes in the structure of liver and intestine with high nitrate ingestion. Further, it was observed that the pathological changes were more with increasing nitrate concentrations. The changes observed were as follows:

Liver

The changes in liver with increasing nitrate ingestion are given in Table 3.3 and Fig. 3.8.

Table 3.3: Different nitrate dosages and histopathological changes in the liver

Nitrate (ppm)	Histopathological changes
100	HP
200	HP, Granular degeneration, mild congestion and inflammation
400	HP, Congestion and inflammation of portal tract
500	HP, Congestion, degeneration and inflammation of portal tract

HP: High power magnification (400X) on microscopic examination

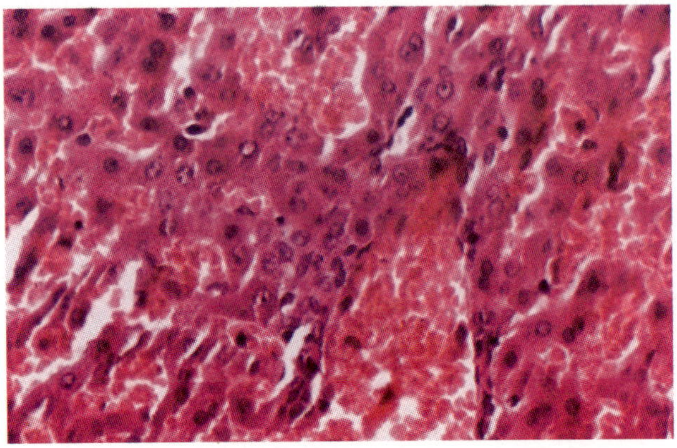

Fig. 3.8. Histopathological changes in the liver at 500 ppm (Hematoxylin and eosin, x 400, Gupta et al. 2001).

Intestine

The histopathological changes in the intestine were not observed at 100 ppm nitrate, but as the nitrate concentration in drinking water increases from 100 ppm, focal collection of inflammatory cells starts at submucosa with slight lymphoid hyperplasia, which was more pronounced as the water nitrate concentration reaches 400 ppm, and at this level, degenerative changes also start (Fig. 3.9).

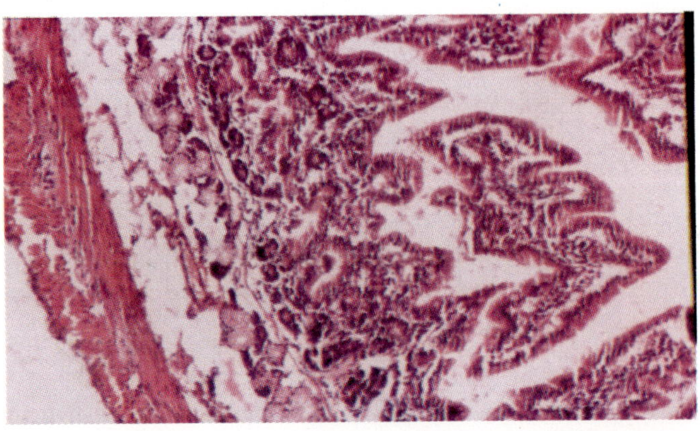

Fig. 3.9. Histopathological changes in the intestine at 400 ppm (LP) (Hematoxylin and eosin, x 100, Gupta et al. 2001).

ABORTIONS

Health effects associated with the ingestion of nitrate-contaminated water include stillbirth, low birth weight, and slow weight gain and even death of the affected animals (Committee on nitrate accumulation, 1972). Spontaneous abortions were also observed in laboratory animals and livestock (Sund et al. 1957; FDA 1972). In 1959, for the first time in humans, spontaneous abortions were reported to have an association with increased methemoglobin levels due to high nitrate ingestion (Muhrer 1959). In subsequent years, some studies (Schmitz 1861; CDC 1996; Fewtrell 2004) found an increased risk of spontaneous abortion or certain birth defects if the mother drank water high in nitrate. Therefore it has been suggested that women who are pregnant or are trying to become pregnant should not consume water containing high nitrate.

BIRTH DEFECTS: MALFORMATIONS

The risk of birth defects due to nitrate exposure is a particular concern because of the fact that it could be due to a single high dose of nitrate early in the pregnancy, which has profound effects on long-term fetal development. Animal studies have indicated that there is a transplacental transfer of N-nitroso compounds to the fetus (Shuval and Gruener 1972) and this fetal exposure can cause cancer later in life (Druckrey 1966).

Extending these findings on human beings, a number of human epidemiology studies were conducted (Knox 1972; Super 1981; Dorsch 1984). A link was found between anencephaly rates and intake of cured meat containing high levels of nitrite (Knox 1972). This study provided the first suggestive evidence in humans that nitrite consumption in food could have adverse impacts on the fetus. A 1984 study (Dorsch et al. 1984) found statistically significant dose response relationships between birth defects of the central nervous system and musculoskeletal system and increasing nitrate concentration of drinking water.

DIABETES

A positive correlation between high nitrate levels in drinking water and increased incidence of type 1 diabetes was observed independently from the length of mother's education, child's or mother's age, place of residence or mother's smoking status (Kostraba 1992; Virtanen et al. 1994; Parslow et al. 1997; Van Maanen et al. 2000). Further studies indicated that consumption of high levels of N-nitroso compounds (NOCs) by human mothers may result in an increased incidence of type 1 diabetes in male offspring (Helgason et al. 1981). In animal studies, NOCs have been reported to be toxic to pancreatic beta cells, providing the rationale for these observations (Wilson et al. 1983). Kostraba (1992) postulated that exposure to nitrate in drinking water causes increased production of free radicals, which may play a role in the etiopathogenesis of insulin-dependent diabetes mellitus (IDDM) (Verge 1994; Mandrup-Poulsen 1996).

THYROID

The thyroid gland contains a iodine trapping transport mechanism, which is accomplished by a membrane protein, the sodium-iodine symporter, which provides sufficient iodine substrate for hormone formation. This trapping mechanism for iodine is shared by other monovalent anions, including pertechnetate, perchlorate, thiocyanate and nitrate, and it is reported that relative potency of perchlorate for inhibiting iodine uptake is 15 and 240 times greater than that of thiocyanate and nitrate, respectively, on a molar concentration basis in serum (Tonacchera M. et al. 2004). Due to inhibition of this trapping mechanism, chronic nitrate exposure causes an inhibition in the accumulation of iodine in the thyroid gland and consequently it may result in thyroid malfunction, causing higher relative risk of goiter (Vladeva et al. 2000) in children, more volume (Van Maanen et al. 1994; Tajtakova M et al. 2006) and weight (Eskiocak 2005; Eskiocak et al. 2005) of thyroid gland in children as well as in adults, and higher frequency of hypoechogenicity (Tajtakova et al. 2006) of thyroid gland. Histomorphological changes reported are retention of lobular architecture, prominent vascular congestion, follicular hyperplasia, a vacuolization and an increase in the colloidal volume (Eskiocak et al. 2005) (Fig. 3.10).

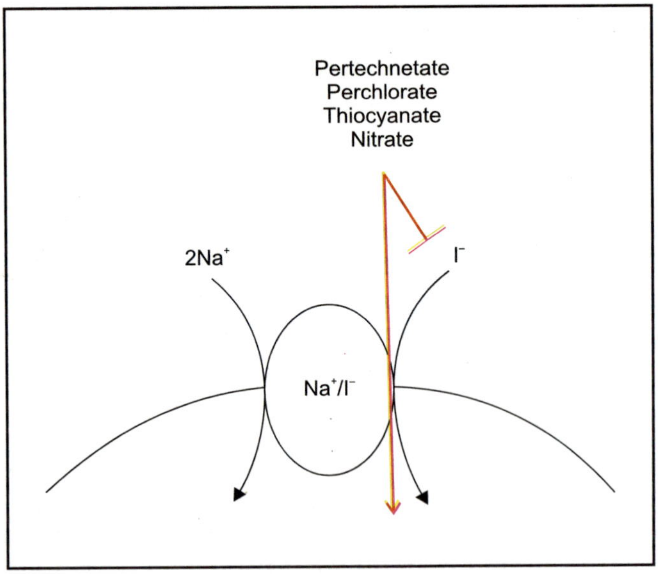

Fig. 3.10. Competition with iodine for sodium-iodine symporter resulting in decreased iodide uptake by thyrocytes.

Several investigators showed that low dose or short-term nitrate intake causes a decrease in thyroid radioiodine uptake (Szokeova et al. 2001), whereas Eskiocak et al. (Eskiocak 2005; Eskiocak et al. 2005) reported that high dose and long-term nitrate exposure results in an increase in the thyroid radioiodine uptake. These findings suggest that the effect of nitrate on thyroid iodine uptake is dose dependent and the inhibition of thyroid iodine uptake may be stronger at higher amounts of nitrate.

In human studies, Van Maanen et al. (1994) reported decreased in TSH levels, whereas Tajtakova et al. (2006) reported an increase in TSH. A decrease in total and free T_3 and T_4 levels at high dose with long-term nitrate exposure (Eskiocak 2005; Eskiocak et al. 2005) has been reported, whereas Szokeova et al (Szokeova et al. 2001) demonstrated that short-term nitrate administration may result in a significantly higher serum level of total T_3.

These findings indicate that short or long-term nitrate exposure may be strongly responsible for the prominent change in thyroid hormone production.

NOx (NITRITE/NITRATE) AND NEPHROTIC SYNDROME

NOx levels in serum obtained from patients with nephritic syndrome showed significantly higher levels than those of healthy controls. Balat et al. (2000) measured plasma and urinary, total NOx in children with minimal change nephritic syndrome (MCNS). In comparison with healthy controls, children with MCNS had increased urinary nitrite excretion. Plasma nitrite levels were high in relapse compared with controls (Balat et al. 2000). Trachtman et al. (1996) also reported that patients with MCNS had increased urinary nitrite excretion regardless of whether the disease was in relapse or remission.

An experiment on rabbit indicated increasing damage to kidney with increasing nitrate concentration in drinking water of the animal (Gupta et al. – unpublished data). Histopathological changes at different nitrate concentration have been depicted in Table 3.4 and Fig. 3.11 given below.

Table 3.4: Different nitrate dosages and histopathological changes in the kidney

Nitrate (ppm)	Histopathological changes
100	HP, Tubular changes, Necrosis (+++)
200	HP, Tubular changes, Eosinophilic changes
400	HP, Tubular changes, Eosinophilic changes, granular changes, increased cellularity

HP: High power magnification (400X) on microscopic examination

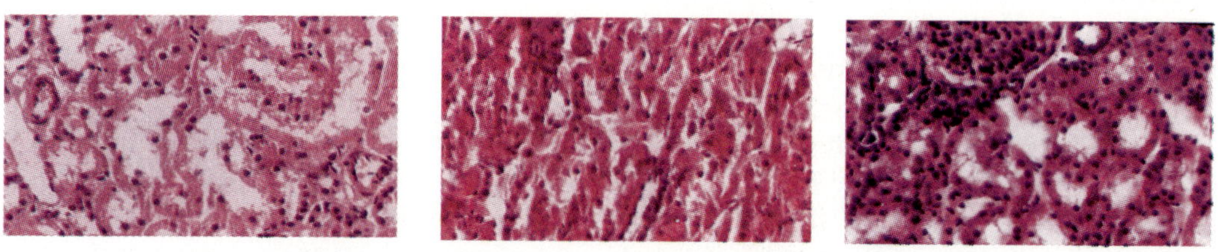

A B C

Fig. 3.11. Histopathological changes in kidneys at different nitrate concentrations. 100, 200 and 400 ppm nitrate intake group at A, B and C, respectively. (Hematoxylin and eosin, x 400, Gupta et al. unpublished data) Matter to present in BOX (1.1)

ADRENAL GLAND

High nitrate ingestion affects the adrenal gland. In addition to the effect of nitrite on the adrenal zona glomerulosa in rats, a study in humans indicated that sodium nitrite (0.5 mg of sodium nitrite per kg of body weight per day, during 9 days) causes a decreased production of adrenal steroids, as reflected by the decreased concentration of 17-hydroxysteroid and 17-ketosteroids in urine (Til et al. 1988; Kuper and Til 1995). Similar results were also found in rabbits (Violante et al. 1973). Although the mechanism is not clear, the effects of nitrite seen in rats seem relevant for the hazard assessment in humans.

IMMUNITY

Few studies (CDFA 1989; Kozliuk et al. 1989; Ustyugova 2002) reported the effect of nitrate/nitrite ingestion on immune system. The effect of nitrate ingestion (Ustyugova 2002) on human immune system indicated that nitrate had no effect on lymphocyte growth, but nitrite decreases the proliferation of lymphocytes. Fibroblast growth remains unaffected. A decreased production of Th1 cytokines (interleukin-2, interferon-gamma, and tumor necrosis factor-beta), which is responsible for resistance to a variety of infectious diseases was noted. No effect on the production of Th2 cytokine interleukin-10, which is responsible for disease susceptibility, was noted. Since nitrate/nitrite shifted the balance from a Th1 to a Th2 response in some individuals, exposure to these compounds may decrease these people's responsiveness to infectious diseases. The levels of nitrate used in this study are relevant to human health because they are present in the liquid portion (non-breastfed) of some 2-month-old infant diets in rural Romania. Animal studies also reported an immune suppression due to high nitrate ingestion (Porter et al. 1999).

AIR POLLUTION

Inhalation of NOx causes a wide variety of health and environmental impacts because of various compounds and derivatives in the family of nitrogen oxides, including nitrogen dioxide, nitric acid, nitrous oxide, nitrates, and nitric oxide. It reacts to form nitrate particles, acid aerosols and contributes to formation of acid rain.

Health effects are related to levels of NOx as well as the duration of exposure (CEOHAATS 1996; Bernstein et al. 2004). Low levels of nitrogen oxides in the air irritate eyes, nose, throat, and lungs leading to cough and shortness of breath, tiredness, and nausea. Breathing high levels of nitrogen oxides can cause rapid burning, spasms, and inflammatory swelling of tissues in the throat and upper respiratory tract. High exposures may lead to pulmonary edema leading to hypoxemia and even death. Industrial exposure to nitrogen dioxide may cause genetic mutations, damage a developing fetus, and decrease fertility in women. Industrial exposure to nitric oxide can cause unconsciousness, vomiting, mental confusion, and damage to the teeth. So far, there is no evidence that nitrogen oxides are potential carcinogens.

A review of studies published in the last decade has shown urban pollution to be environmental cardiovascular risk factors (Maitre et al. 2006). This link was significant for NOx and PM10. A study of short-term effects of nitrogen dioxide on total, cardiovascular and respiratory mortality in 30 European cities found significant association between the two (Samoli et al. 2006). Significant associations of daily changes in particle concentrations, nitrogen dioxide and carbon monoxide were found with hospitalization for respiratory diseases (COPD, pneumonia, asthma) and cardiovascular diseases (Hinwood et al. 2006) . Exposure to indoor NO_2 at levels well below the Environmental Protection Agency outdoor standard (53 ppb) were associated with respiratory symptoms among children with asthma. Each 20-ppb increase in NO_2 increased both the likelihood of any wheeze or chest tightness and days of wheeze or chest tightness (Belanger et al. 2006).

The EPA has established that the average concentration of nitrogen dioxide in ambient air in a calendar year should not exceed 0.053 parts of nitrogen dioxide per million parts of air (0.053 ppm). The Occupational Safety and Health Administration (OSHA) have set a limit of 25 ppm of nitric oxide in workplace air during an 8-hour workday, 40-hour work week. OSHA has also set a 15-minute exposure limit of 5 ppm for nitrogen dioxide in workplace air.

ANIMAL EXPERIMENTS

Clinical manifestations of nitrate experiments on different systems are as follows:

Central Nervous System

The results of experiments on mice indicated a nitrate-induced deviation in behavioral development, and impairment in learning behavior, particularly of the discriminative type (Markel et al. 1989).

Gastrointestinal System

In a 19-month study on mice, the effects of nitrate on gastric epithelium (Ptashekas 1990) indicated atypical changes in the gastric epithelium.

Cardiovascular System

Animal studies (Gruener et al. 1970; Shuval et al. 1972) reported increased cardiac muscle contraction, which correlated directly to methHb levels. At 10–15% methHb, ECG shows shortening of Q-T interval and reduction in T wave.

Embryo Toxicity

Experiments on rats showed anemia, increased meth Hb, and increased mortality in offspring. (RIVM 1989; WHO 1996).

Genotoxicity

In an animal experiment on mice, an increase in chromosome aberrations was found in a group of animals, who were consuming nitrate more than 707 mg/kg BW. (Luca et al. 1985).

Reproductive Behavior

Reproduction in female guinea pig (RIVM 1989; WHO 1996) was grossly impaired by high nitrate ingestion (30,000 mg of potassium nitrate per liter).

Mutagenicity and Related End-Points

Nitrite is mutagenic. It causes morphological transformations in *in vitro* systems. (RIVM 1989; WHO 1996).

Serum Somatomedin Activity

A decrease in serum somatomedin activity due to nitrate administration was also observed (Jahreis et al. 1987).

TREATMENT AND PREVENTION

Treatment

Supplementation of ascorbic acid, methionine, alpha-tocopherol, and methylene blue has been found to be effective in overcoming the problem of nitrate toxicity. These substances are effective in inhibiting nitrosation if present in gastric juices at all times in significant concentrations.

Role of Antioxidants

As ROS are derived from a number of sources including nitrates, it was deemed feasible to readjust the balance of ROS production and detoxification by supplementing antioxidants. How to prevent or treat this oxidative stress? (Armitage et al. 2009).

However, clinical trials testing this hypothesis showed little benefits in reducing respiratory, GIT, cardiovascular etc. events or mortality; some of the treatments even caused harm (Dotan et al. 2009).

Indeed, *in vivo* detoxification of increased ROS by antioxidant supplementation may be impossible due to the poor bioavailability of antioxidants at the correct subcellular localization and at the optimal time point. Antioxidants may also promote new radical chain reactions initiated by their oxidized forms.

Importantly, the global removal of ROS may result in unwanted effects since ROS also regulate important physiological functions. Therefore, the current strategy is to identify and target the major sources of ROS, including NADPH oxidases

(NOx). A major constraint in NADPH oxidase research and translation is the lack of specific inhibitors.

Prevention

At environmental level

Avoid water pollution

- Minimize the contamination of water supplies by nitrates originating from agricultural practice: Avoid inadvertent and excessive use of nitrogenous fertilizers
- Avoid water pollution
- Avoid the habit of open air defecation
- Avoid the stagnation of waste water around the source of water
- Avoid the sewage disposal directly to groundwater table

Removal of nitrate from drinking water—Denitrification

The removal of nitrate from water is costly and difficult to implement and maintain both at domestic as well as community level. The following are the known processes of denitrification:

a. Physiochemical process—The commonly used processes under this category are (a) Reverse osmosis (b) Ion exchange (c) Electrodialysis
b. Biological Treatments—This process is important since it bears similarity with the metabolism of nitrates in living organisms. (Details of these processes have been depicted in annexure II)

At human level

- Breastfeeding only, at least upto the age of 4 months.
- If top feed is necessary, then preferably use cow milk (of course unadulterated) or dry milk preparations reconstituted with water containing low level of nitrate.
- Avoid the use of high nitrate containing food for weaning.
- Avoid the use of nitrate and nitrite as preservative to the minimum extent possible especially to cured and canned meats and fish.
- Use of drinking water with low level of nitrates especially by pregnant mothers.
- Avoiding WHO ORS preparation with locally available high nitrate water during diarrhea.
- Avoid long-term use of anti-acid secretory agents especially in pregnant mothers and children. In case it is necessary to use these drugs, they should be used cautiously and preferably with the use of antioxidants.

Reducing nitrate levels in vegetables

The following hints will reduce nitrates in vegetables. The problem is that some of them will reduce essential nutrients such as vitamin C as well.

- Washing and cooking in water (nitrates are soluble in water).
- Peeling, e.g. in potatoes because the nitrates are concentrated in the skin and just below (again, this is part of fail-safe eating).
- Discarding: in leafy vegetables such as lettuce, discarding the stem and mid-rib can decrease up to 40 percent of the nitrates
- A study of French fries showed that peeling reduced the levels by 30 percent; preheating and cutting reduced the nitrate content by a further 20 percent and blanching by 30 percent. After final frying only about five per cent of the original nitrate content remained. (Reheating cold cooked vegetables has been shown not to increase nitrate levels as previously thought).
- For homegrown vegetables, choose low nitrate varieties (iceberg lettuce is one of the lowest and is fail-safe); use slower releasing nitrogen sources such as manure and compost; avoid fertilizing just before harvest; grow vegetables outside if possible (lower light due to plastic or glass covering contributes to higher nitrates); harvest in full sun.
- A number of studies have suggested a protective effect of vitamin C and other nutrients against either nitrates or particular cancers. Some experts think that the naturally occurring nutrients in vegetables are enough to protect against effects of nitrates. The WCRF warns: do not use high-dose supplements to protect against cancer. Research shows that high-dose nutrient supplements can affect our risk of cancer, so it is best to opt for a balanced diet without supplements.

CONCLUSION

Nitrate and nitrite are natural ions that are a part of nitrogen cycle. Naturally occurring nitrate levels in surface and groundwater are generally a few milligrams per liter. In many groundwaters, an increase in nitrate levels has been observed due to water percolating through nitrate rich rocks and owing to the farming practices of using chemical fertilizers.

For human consumption, WHO report, 2004 permits up to 50 mg/L but for short-term exposure only, whereas BIS-10500 permits 45 mg/L (as NO_3) as desirable limit and 100 mg of NO_3 per liter as maximum permissible limit in the absence of alternate source.

Not much literature on human studies is available on nitrate toxicity except reports documenting methemoglobinemia in infants due to high nitrate ingestion. Apart from methemoglobinemia, few studies indicated nitrate as a cause of cancer, but it is still controversial and no firm conclusions have been drawn so far. The other effects observed were increased infant mortality, abortions, birth defects, recurrent diarrhea, recurrent stomatitis, early onset of hypertension, ischemia, CVD, histopathological changes in cardiac muscles, alveoli of lungs and adrenal glands, recurrent respiratory tract infection in children, nephrotic syndrome in children, hypothyroidism and diabetes. Recent studies have indicated that high nitrate ingestion adversely affects the immune system of the body. Recently an adaptation system to nitrate ingestion has also been reported. This adaptation to an enzyme cytochrome b_5 reductase has been shown to be protective to human beings, but only to a limited extent.

More detailed epidemiological health related studies covering a large sample are required to provide insight to the nitrate toxicity in human beings. It is possible that a detailed study on pathophysiology of nitrate metabolism and its effect on human beings would yield a better understanding of the various diseases caused by nitrates, and their prevention and treatment.

Since it is very difficult and costly to remove nitrate from water because it is chemically non-reactive in dilute aqueous solutions, it is recommended that a simple change in habits and adoption of simple preventive measures may be urgently introduced in government campaign at least in high nitrate belts. Indiscriminate use of nitrogenous fertilizers should be avoided. The most important strategy is to promote breastfeeding up to the age of at least 6 months.

ANNEXURE – I

Nitrates and Nitrites in Vegetables (mg/kg)

Nitrate

Food	Low	Average	High	Year	Source*
Beans	6	392	810	2008	EFSA
Beetroot	110	1370	3670	2008	EFSA
Brussels sprouts	1	24	100	2008	EFSA
Cabbage	47	311	833	2008	EFSA
Carrots	21	296	1574	2008	EFSA
Cauliflower	7	148	148	2008	EFSA
Celery	18	1103	3319	2008	EFSA
Garlic	8	69	161	2008	EFSA
Green Beans	9	323	735	2008	EFSA
Leek	5	345	975	2008	EFSA
Lettuce iceberg	210	875	1537	2008	EFSA
Parsnip	2	16	83	2008	EFSA
Peas	1	30	100	2008	EFSA
Potatoes	10	168	340	2008	EFSA
Pumpkin	8	894	4617	2008	EFSA
Rhubarb	28	2943	6550	2008	EFSA
Spinach	64	1066	3048	2008	EFSA

Spinach organic	2138	2005	FSA
Lettuce glasshouse	5700	2004	FSA
Lettuce rucola	4800	2008	EFSA
Lettuce curly	3263	2005	FSA
Lettuce iceberg u/cover	2500	2004	FSA
Lettuce organic	1115	2005	FSA
Lettuce	875	2008	EFSA
Lettuce organic	596	1982	SA

http://www.efsa.europa.eu/cs/BlobServer/Scientific_Opinion/contam_ej_689_nitrate_en.pdf
The EFSA maximum permitted level for nitrates in lettuce is 4,500 mg/kg

ANNEXURE – II

BIOLOGICAL DENITRIFICATION

The biological reduction of nitrate to nitrite and subsequently to dinitrogen gas requires a suitable electron donor. The electron donor is usually an organic molecule, and methanol is the most commonly used carbon source. Equations 1 and 2 represent the reduction of nitrate to nitrite and nitrite to nitrogen gas respectively, and equation 3 represents the overall reaction using methanol as the electron donor (McCarty et al. 1969).

$$NO_3^- + 0.33\ CH_3OH \longrightarrow NO_2^- + 0.67\ H_2O + 0.33\ CO_2 \quad (1)$$
$$NO_2^- + 0.5\ CH_3OH \longrightarrow 0.5\ N_2 + 0.5\ H_2O + 0.5\ CO_2 + OH^- \quad (2)$$

and overall:

$$NO_3^- + 0.833\ CH_3OH \longrightarrow 0.5\ N_2 + 1.16\ H_2O + 0.833\ CO_2 + OH^- \quad (3)$$

Other organic carbon sources such as ethyl alcohol sucrose, acetone, brewery waste, chemical process waste, corn starch waste, molasses, wharf, sulfide liquor and winery residue, etc. are taken as inexpensive carbon substrates.

Autotrophic denitrification has been studied using hydrogen or various sulfur compounds. Autotrophic denitrifying bacteria use molecular hydrogen or other inorganic compounds as reductants and carbon dioxide as the source of carbon as shown in equation below:

$$5\ H_2 + 2H^+ + 2\ NO_3 \longrightarrow N_2 + 6\ H_2\ O \quad (4)$$

Some researchers have evaluated the use of reduced sulfur compounds such as sulfide and thiosulfate for the denitrification of water and domestic and industrial wastewater (Sutton et al. 1979; Amminudin and Nicholas 1973). Sulfate is a by-product of denitrification using sulfur compounds as given in the following equations.

Thiosulfate (Claus and Kutzner 1985)

$$5\,S_2O_3^{2-} + 8\,NO_3^- + H_2O \longrightarrow 4\,N_2 + 10\,SO_4^{2-} + 2H^+ \qquad (5)$$

Sulfide (Barrenstein et al. 1986)

$$5\,S^{2-} + 8\,NO_3^- + 8\,H^+ \longrightarrow 5\,SO_4^{2-} + 4\,N_2 + 4\,H_2O \qquad (6)$$

Though oxygen inhibits denitrification, there have been periodic reports of aerobic denitrification (Marshall et al. 1953; Krul 1976). Recently under controlled conditions in homogeneously suspended bacterial cultures at D.O. concentrations ranging from 10% to twice air saturation, persistent denitrification has been reported by many workers (Meiberg et al. 1980; Gupta and Kumar 1999). Robertson et al. (1988) not only detected the presence of appropriate enzyme, but also demonstrated the production of nitrogen containing gases from nitrate by *Thiosphaera pantotropha* (isolated from desulphurizing, denitrifying wastewater treatment systems) at dissolved oxygen concentrations up to 90% of air saturation.

Biological denitrification has a limited role in the treatment of drinking water because introduction of carbon source and bacteria to water for denitrification would increase the post-treatment cost significantly. Several investigators have evaluated the injection of various substrates and nutrients into aquifers in an effort to simulate *in situ* denitrification. Liquid substrates such as acetic acid, ethanol and treated wastewater have been used and gaseous substances have been evaluated (Kruithof et al. 1988; Soares et al. 1988; Gayle et al. 1989). Application of *in situ* groundwater denitrification depends on the prevention of well clogging and biomass accumulation is thought to be the main cause of *in situ* clogging. In shallow aquifers accumulation of gas may represent a major contributor to the clogging of wells. It is important to take certain precautions while designing an *in situ* denitrification scheme, e.g., complete utilization of carbon sources should be ensured since no organic carbon is allowed to be present in drinking water; clogging of aquifer must be anticipated particularly in soil that provide small pore size; release of N_2 gas must be ensured; movement of bacteria in porous media should be controlled by giving due consideration to factors like ratio of cell size and pore size, shape of microorganisms, flow velocity, injection concentration of bacteria, etc. A proper balance of the above factors should be made to ensure uniform dispersion of bacteria in sub-soil. Biological denitrification has been extensively used for the removal of nitrates from domestic and industrial wastewaters, which can help prevent nitrate contamination of groundwater, however, only certain advancements in this process have been summarized in the subsequent paragraph.

Gupta (1997) gave a comprehensive review of the literature pertaining to enzyme system of *Thiosphaera pantotropha*, which is responsible for the nitrification and denitrification properties and its potential applications for wastewater treatment. Many laboratory studies were carried out in different suspended growth and fixed film systems to bring out low cost options for a simultaneous removal of carbon and nitrogen from different synthetic samples simulating domestic and industrial waste-waters (Gupta and Gupta 1999, 2001; Gupta et al. 1994).

REFERENCES

Allison, M.J. and Reddy, C.A. (1984). Adaptation of gastrointestinal bacteria in response to the changes in dietary oxalate and nitrate. *In:* Klug, M.J. and Reddy, C.A. (Eds.). Current perspectives in microbial ecology, *Am. Soc. Microbio.* pp. 248-256, Washington D.C.

Aminuddin, M. and Nicholas, D.J. (1973). Sulphide oxidation linked to the reduction of nitrate and nitrite in *Thiobacillus denitrificans. Biochim. Biophys. Acta* **325:** 81-93.

Anjana, Umar S., Iqbal, M. and Abrol, Y.P. (2007a). Are nitrate concentrations in leafy vegetables within safe limits? *Curr. Sci.* **92(3):** 355-359.

Anjana, Umar S. and Iqbal, M. (2007b). Nitrate accumulation in plants, factors affecting the process, and human health implications. *A review Agron. Sust. Develop* **27:** 45-57.

Armijo, R. and Coulson, A.H. (1975). Epidemiology of stomach cancer in Chile—the role of nitrogen fertilizers. *Int. J. Epidemiol.* **4:** 301-309.

Armitage, M.E., Wingler, K., Schmidt, H.H. and La, M. (2009). Translating the oxidative stress hypothesis into the clinic: NOX versus NOS. *J. Mol. Med.* **87:** 1071-1076.

Balat, A., Cekmen, M., Yürekli, M., Gülcan, H., Kutlu, O., Türköz, Y. and Yologlu, S. (2000). Adrenomedullin and nitrite levels in children with minimal change nephrotic syndrome. *Pediatr Nephrol* **15:** 70-73.

Barrenstein, A., Kramer, U. and Obermann, P. (1986). Underground treatment of nitrate rich groundwater by infiltration with treated wastewater or methane-rich natural gas. DVGW- Schriftreihe, Wasser, Frankfurt, West Germany **116:** 99-116.

Bartholomew, B. and Hill, M.J. (1984). The pharmacology of dietary nitrate and the origin of urinary nitrate. *Food Chem. Toxicol.* **22:** 789-795.

Belanger, K., Gent, J.F., Triche, E.W., Bracken, M.B. and Leaderer, B.P. (2006). Association of indoor nitrogen dioxide exposure with respiratory symptoms in children with asthma. *Am. J. Respir Crit. Care Med.* **173:** 297-303.

Benz, E.J. (2001). Hemoglobenopathies: Hemoglobins with altered oxygen affinity. *In:* Braunwald, E., Fauci, A.S., Kasper, D.L., Hauser, S.L., Longo, D.L. and Jameson, J.L. (Eds.). Harrison's Principles of Internal Medicine 1, 15th Ed., pp. 1524. McGraw–Hill, New York.

Berger, H.M., Moison, R.M.W. and Van Zoeren-Grobben, D. (1997). Pathophysiology of respiratory distress syndrome. *In:* David, T.J. (Ed.). Recent advances in pediatrics, vol: 14, New Delhi, Churchill livingstone, pp. 117-119, New Delhi.

Bernstein, J.A., Alexis, N., Barnes, C., Bernstein, I.L., Bernstein, J.A., Nel, A., Peden, D., Diaz-Sanchez, D., Tarlo, S.M. and Williams, P.B. (2004). Health effects of air pollution. *J. Allergy Clin. Immunol.* **114:** 1116-1123.

Bodansky, O. (1951). Methemoglobinemia and methemoglobin-producing compounds. *Phamacol. Rev.* **3:** 144-196.

Calif. Dept. of Food and Agric. (CDFA), Nitrate Working Group. (1989). Nitrate and Agriculture in California.

CDC (Centers for Disease Control and Prevention), (1996). Spontaneous abortions possibly related to ingestion of nitrate-contaminated well water–LaGrange County, Indiana, 1991–1994. *Morb. Mortal. Wkly. Rep.* **45:** 569-572.

CEOHAATS (Committee of the Environmental and Occupational Health Assembly of the American Thoracic Society), (1996). Health effects of outdoor air pollution. *Am. J. Respir. Crit. Care Med.* **153:** 3-50.

Choi, B.C. (1985). N-nitroso compounds and human cancer: a molecular epidemiologic approach. *Am. J. Epidemiol.* **121:** 737-743.

Claus, G. and Kutzner, H.J. (1985). Autotrophic denitrification by *Thiobacillus denitrificans* in a packed bed reactor, *Appl. Microbiol. Biotechnol.* **22:** 289-296.

Colbers, E.P.H., Hegger, C., Kortboyer, J.M. and Meulenbelt, J. (1996). A pilot study to investigate nitrate and nitrite kinetics in healthy volunteers with both normal and artificially increased gastric pH after sodium nitrate ingestion. Report No. 235802001 of the National Institute of Public Health and the Environment (RIVM), Bilthoven, pp. 1-62. Netherlands.

Cole, J.A. and Brown, C.M. (1980). Nitrate reduction to ammonia by fermentative bacteria: a short circuit in the biological nitrogen cycle. *FEMS Microbiol. Lett.* **7:** 65-72.

Colvin, C., Tredoux, G., Clarke, S., Le Maitre, D., Englebrecht, P. and Maherry, A. (2008). Integration goes underground: A review of groundwater research in support of sustainable development in South Africa. Science real and relevant: 2nd CSIR International Convention Centre, Pretoria, 17th and 18th November 2008, pp. 10.

Comly, H.H. (1945). Cyanosis in infants caused by nitrates in well water. *JAMA* **129:** 112-116.

Committee on Nitrate Accumulation, Ag. Board, Div. of Biology & Agriculture, National Research Council, (1972). Hazards of nitrate, nitrite, and nitrosoamines to man and livestock. *In:* Accumulation of Nitrate, National Academy of Sciences, pp. 46-75, Washington D.C.

Cornblath, M. and Hartmann, A.F. (1948). Methaemoglobinemia in young infants. *J. Pediat.* **33:** 421-425.

Corre, W.J. and Breimer, T. (1979). Nitrate and nitrite in vegetables. Literature Survey nr. 39, Center for Agricultural Publishing Documentation, Wageningen.

Cortas, N.K. and Wakid, N.W. (1991). Pharmacokinetic aspects of inorganic nitrate ingestion in man. *Pharmacol. Toxicol.* **68:** 92-195.

Craven, P.A. and DeRubertis, F.R. (1978). Restoration of the responsiveness of purified guanylate cyclase to nitrosoguanidine, nitric oxide, and related activators by heme and hemeproteins. Evidence for involvement of paramagnetic nitrosyl-heme complex in enzyme activation. *J. Biol. Chem.* **253:** 8433-8443.

Cuello, C., Correa, P., Haenszel, W., Cordillo, G., Brown, C., Archer, M. and Tannenbaum, S. (1976). Gastric cancer in Colombia I. Cancer risk and suspect environmental agents. *J. Natl. Cancer Inst.* **57:** 1015-1020.

Dallinga, J.W., Pachen, D.M., Lousberg, A.H., van Geel, J.A., Houben, G.M., Stockbrügger, R.W., van Maanen, J.M. and Kleinjans, J.C. (1998). Volatile N-nitrosamines in gastric juice of patients with various conditions of gastrointestinal tract determined by gas chromatography–mass spectrometry and related to intragastric pH and nitrate and nitrite levels. *Cancer Lett.* **124:** 119-125.

De Roos, A.J., Ward, M.H., Lynch, C.F. and Cantor, K.P. (2003). Nitrate in public water supplies and the risk of colon and rectum cancers. *Epidemiology* **14:** 640-649.

Dolby, J.M., Webster, A.D., Borriello, S.P., Barclay, F.E., Bartholomew, B.A. and Hill, M.J. (1984). Bacterial colonization and nitrite concentration in the achlorhydric stomachs of patients with primary hypogammaglobulinaemia or classical pernicious anaemia. *Scand J. Gastroenterol* **19:** 105-110.

Dorsch, M.M., Scragg, R.K., McMichael, A.J., Baghurst, P.A. and Dyer, K.F. (1984). Congenital malformations and maternal drinking water supply in rural South Australia: a case-control study. *Am. J. Epidemiol.* **119:** 473-486.

Dotan, Y., Pinchuk, I., Lichtenberg, D. and Leshno, M. (2009). Decision analysis supports the paradigm that indiscriminate supplementation of vitamin E does more harm than good. *Arterioscler Thromb Vasc. Biol.* **29:** 1304-1309.

Druckrey, H., Ivankovic, S. and Preussmann, R. (1966). Teratogenic and carcinogenic effects in the offspring after single injection of ethylnitrosourea in pregnant rats. *Nature* **210:** 1378-1379.

Duguid, J.B. (1960). The thrombogenic hypothesis and its implications. *Postgrad. Med. J.* **36:** 226-229.

Eisenbrand, G., Spiegelhalder, B. and Preussmann, R. (1980). Nitrate and nitrite in saliva. *Oncology* **37:** 227-231.

El Nahas, S.M., Globus, M. and Vethamany-Globus, S. (1984). Chromosomal aberrations induced by sodium nitrite in bone marrow of adult rats and liver cells of transplacentally exposed embryos. *J. Toxicol. Environ. Health* **13:** 643-647.

Ellen, G. and Schuller, P.L. (1983). Nitrate, origin of continuous anxiety. *In:* Preusmann, R. (Ed.). Das Nitrosamin-Problem. Deutsche Forschungsgemeinschaft, Verlag Chemie GmbH, Weinheim, pp. 97-134.

Eskiocak, S. (1995). [The effects of administration chronic nitrate by drinking water on thyroid hormones, thyroidal radioiodine uptake and mass]. Thesis in Biochemistry Speciality. Samsun, Turkey.

Eskiocak, S., Dundar, C., Basoglu, T. and Altaner, S. (2005). The effects of taking chronic nitrate by drinking water on thyroid functions and morphology. *Clin. Exp. Med.* **5:** 66-71.

Farinati, F., Della Libera, G., Cardin, R., Molari, A., Plebani, M., Rugge, M., Di Mario, F. and Naccarato, R. (1996). Gastric antioxidant, nitrites, and mucosal lipoperoxidation in chronic gastritis and *Helicobacter pylori* infection. *J. Clin. Gastroenterol.* **22:** 275-281.

Farinati, F., Lima, V., Naccarato, R. and Garro, A. (1989). Mutagenic activity in gastric juice and urine of subjects with chronic atrophic gastritis, gastric epithelial dysplasia and gastric cancer. *Cancer Lett.* **48:** 169-175.

Fassett, D.W. (1973). Nitrates and nitrites. In Toxicants occurring naturally in foods. Washington DC, *Natl. Acad. Sci.* 7-25.

Fewtrell, L. (2004). Drinking water nitrate, methemoglobinemia, and global burden of disease: a discussion. *Environ. Health Perspect* **112:** 1371-1374.

Food and Drug Administration, (1972). Teratologic evaluation of FDA 71-7 (sodium nitrate). Washington, DC: US Department of Health and Human Services, Public Health Service, Food and Drug Administration, Publication No. PB 221775.

Fraser, P., Chilvers, C., Beral, V. and Hill, M.J. (1980). Nitrate and human cancer: a review of the evidence. *Int. J. Epidemiol.* **9:** 3-11.

Gangolli, S.D., van den Brandt, P.A., Feron, V.J., Janzowsky, C., Koeman, J.H., Speijers, G.J.A., Speigelhalder, B., Walker, R. and Wishnok, J.S. (1994). Nitrate, nitrite and N-nitroso compounds. *Eur. J. Pharmacol.* **292:** 1-38.

Garg, H.K. and Bryan, N.S. (2009). Dietary sources of nitrite as a modulator of ischemia/ reperfusion injury. *Kidney Int.* **75:** 1140–1144.

Gayle, B.P., Boardman, G.D., Sherrard, J.H. and Benoit, R.E. (1989). Biological dentirification of water. *J. Environ. Eng.* (New York) **115:** 930.

Gilli, G., Corrao, G. and Favilli, S. (1984). Concentrations of nitrites in drinking water and incidence of gastric carcinomas: First descriptive study of the Piemonte Region, Italy. *Sci. Total Environ.* **34:** 35-48.

Green, L.C., Ruiz de Luzuriaga, K., Wagner, D.A., Rand, W., Istfan, N., Young, V.R. and Tannenbaum, S.R. (1981). Nitrate biosynthesis in man. *Proc. Natl. Acad. Sci. USA.* **78:** 7764-7768.

Gruener, N. and Shuval, H.I. (1970). Health aspects of nitrates in drinking water. *In:* Developments in Water Quality Research, Ann Arbor Humphrey Science Publishers, *Ann. Arbor.*, pp. 89-106, London.

Gupta, A.B. and Gupta, S.K. (1998). Removal of organics and nitrogen from sewage in an aerobic biofilm containing *Thiosphaera pantotropha. Proc. Asian Conference on water and wastewater management*, Tehran, March, 2-4, pp. 562-580.

Gupta, A.B. and Gupta, S.K. (1999). Simultaneous carbon and nitrogen removal in a mixed culture aerobic RBD biofilm. *Water Research.* **33:** 555-561.

Gupta, A.B. and Gupta, S.K. (2001). Simultaneous carbon and nitrogen removal from high strength domestic wastewater in an aerobic RBC biofilm. *Water Res.* 35: 1714-1722.

Gupta, A.B. and Kumar, S. (1999). Aerobic denitrification: myth or reality? *Ind. J. Environ. Sci.* **3:** 85-90.

Gupta, A.B. (1997). *Thiosphaera Pantotropha:* a sulphur bacterium capable of simultaneous heterotrophic nitrification and aerobic denitrification. *Enz. Microbial. Technol.* **21:** 589-595.

Gupta, N. and Chandrasekharan, H.R. (1999). Status of nitrate contamination in groundwater near Delhi: A case study. *In:* Managing Risks of Nitrates to Humans and the Environment (Special Publication), Wilson, W.S., Ball, A.S. and Hinton, R.H. (Eds.). *Royal Society of Chemistry* (spec. publ.) **237:** 240-246.

Gupta, S.K., Fitzgerald, J.F., Chong, S.K.F., Croffie, J.M. and Garcia, J.G. (1998). Expression of inducible nitric oxide synthase (iNOS) mRNA in inflamed esophageal and colonic mucosa in pediatric population. *Am. J. Gastroenterol.* **93:** 795-798.

Gupta, S.K., Gupta, R.C., Gupta, A.B., Seth, A.K., Bassin, J.K., Gupta, A. and Sharma, M.L. (2001). Recurrent diarrhea in areas with high nitrate in drinking water. *Arch Environ. Health* **56:** 369-373.

Gupta, S.K., Gupta, R.C., Seth, A.K., Gupta, A.B., Bassin, J.K. and Gupta, A. (1999b). Adaptation of cytochrome-b5 reductase activity and methaemoglobinaemia in areas with high nitrate concentration in drinking water. *Bull. World Health Organ.* **77:** 749-753.

Gupta, S.K., Gupta, R.C., Seth, A.K., Gupta, A.B., Bassin, J.K. and Gupta, A. (2000). Methaemoglobinaemia in areas with high nitrate concentration in drinking water. *Natl. Med. J. India* **13:** 58-61.

Gupta, S.K., Gupta, R.C., Seth, A.K., Gupta, A.B., Bassin, J.K., Gupta, D.K. and Sharma, S. (1999a). Epidemiological evaluation of recurrent stomatitis, nitrates in drinking water and cytochrome b5 reductase activity. *Am. J. Gastroenterol.* **94:** 1808-1812.

Gupta, S.K., Gupta, R.C., Seth, A.K., Gupta, A.B., Sharma, M.L. and Gupta, A. (1999c). Toxicological effects of nitrate ingestion on cardiorespiratory tissues in rabbit. *S. Asian J. Prev. Cardio.* **2:** 101-105.

Gupta, S.K., Raja, S. and Gupta, A.B. (1994). Simultaneous nitrification-denitrification in a rotating biological contactor. *Environ. Technol.* **15:** 145-153.

Hegesh, E. and Shiloah, J. (1982). Blood nitrates and infantile methemoglobinemia. *Clin. Chim. Acta.* **125:** 107-115.

Helgason, T. and Jonasson, M.R. (1981). Evidence for a food additive as a cause of ketosis-prone diabetes. *Lancet* **2:** 716-720.

Hill, M.J. Hawksworth, G. and Tattersall, G. (1973). Bacteria, nitrosamines and cancer of the stomach. *Br. J. Cancer* **28:** 562-567.

Hinwood, A.L., De Klerk, N., Rodriguez, C., Jacoby, P., Runnion, T., Rye, P., Landau, L., Murray, F., Feldwick, M. and Spickett, J. (2006). The relationship between changes in daily air pollution and hospitalizations in Perth, Australia 1992-1998: a case-crossover study. *Int. J. Environ. Health Res.* **16:** 27-46.

Hsia, C.C. (1998). Respiratory function of hemoglobin. *N. Engl. J. Med.* **338:** 239-248.

Human Digestive System Encyclopedia Britannica, (2007). http://www.britannica.com/eb/article-45361/human-digestive-system#294193.hook. Retrieved 2007-08-22.

IARC, (1978). IARC Monographs on the Evaluation of Carcinogenic Risks of Chemicals to Man. Volume 17: Some N-Nitroso Compounds. World Health Organization.

International Agency for Research on Cancer, (2004). Tobacco Smoke and Involuntary Smoking, IARC Monographs (International Agency for Research on Cancer, Lyon, (France), Vol. 83.

IS 10500, (1995). Drinking water specification. Bureau of Indian standards, New Delhi, pp. 3.

Jaffé, E.R. (1981). Methemoglobinemia. *Clin. Haematol.* **10:** 99-122.

Jahreis, G., Schöne, F., Lüdke, H. and Hesse, V. (1987). Growth impairment caused by dietary nitrate intake regulated via hypothyroidism and decreased somatomedin. *Endocrinol. Exp.* **21:** 171-180.

Jaskiewicz, K., Van Helden, P.D., Wiid, I.J., Steenkamp, H.J. and Van Wyk, M.J. (1990). Chronic atrophic gastritis, gastric pH, nitrites and micronutrient levels in a population at risk for gastric carcinoma. *Anticancer Res.* **10:** 833-836.

Knotek, Z. and Schmidt, P. (1964). Pathogenesis, incidence and possibilities of preventing alimentary nitrate methemoglobinemia in infants. *Pediatrics* **34:** 78-83.

Knox, E.G. (1972). Anencephalus and dietary intakes. *Br. J. Prev. Soc. Med.* **26:** 219-223.

Kostraba, J.N., Gay, E.C., Rewers, M. and Hamman, R.F. (1992). Nitrate levels in community drinking waters and risk of IDDM. An ecological analysis. *Diabetes Care* **15:** 1505-1508.

Kozliuk, A.S., Kushnir, G.V., Anisimova, L.A., Shroit, I.G. and Opopol, N.I. (1989). Immunological status of children living in a region with an increased level of nitrates in the drinking water. *Gig. Sanit.* **54:** 19-22.

Kruithof, J.C., van Bennckom, C.A., Dierx, H.A.L., Hijnen, W.A.M., van Passen, J.A.M. and Schippers, J.C. (1988). Nitrate removal from groundwater by sulphur/limestone filteration. *Water Supply* **6:** 207-217.

Krul, J.M. and Veeningen, R. (1977). The synthesis of dissimilatory nitrate reductase under aerobic conditions in a number of denitrifying bacteria isolated from activated sludge and drinking water. *Water Res.* **11:** 39-43.

Krul, J.M. (1976). Dissimilatory nitrate and nitrite reduction under aerobic conditions by aerobically and anaerobically grown *Alcaligenes* sp. and by activated sludge. *J. Appl. Bacteriol.* **40:** 245-260.

Kumar, S., Gupta, A.B. and Gupta, S. (2002). Need for revision of nitrate standards for drinking water: a case study of Rajasthan. *Indian J. Environ. Health* **44:** 168-172.

Kuper, F. and Til, H.P. (1995). Subchronic toxicity experiments with potassium nitrite in rats. In: Health aspects of nitrate and its metabolites (particularly nitrite). Proceedings of an international workshop, Bilthoven (Netherlands), 8–10 November 1994. Council of Europe Press, pp. 195-212. Strasbourg.

Lee, C., Weiss, R. and Horvath, D.J. (1970). Effects of nitrogen fertilization on the thyroid function of rats fed 40 percent orchard grass diets. *J. Nutr.* **100:** 1121-1126.

Li, H., Duncan, C., Townend, J., Killham, K., Smith, L.M., Johnston, P., Dykhuizen, R., Kelly, D., Golden, M., Benjamin, N. and Leifert, C. (1997). Nitrate reducing bacteria on rat tongues. *Appl. Environ. Microbiol.* **63:** 924-930.

Lowestein, C.J., Dinerman, J.L. and Snyder, S.H. (1994). Nitric oxide: a physiological messenger. *Ann. Intern. Med.* **120:** 227-237.

Luca, D., Raileanu, L., Luca, V. and Duda, R. (1985). Chromosomal aberrations and micronuclei induced in rat and mouse bone marrow cells by sodium nitrate. *Mutat Res.* **155:** 121-125.

Maitre, A., Bonneterre, V., Huillard, L., Sabatier, P. and de Gaudemaris, R. (2006). Impact of urban atmospheric pollution on coronary disease. *Eur. Heart J.* **27:** 2275-2284.

Malberg, J.W., Savage, E.P. and Osteryoung, J. (1978). Nitrates in drinking water and the early onset of hypertension. *Environ. Pollut.* **15:** 155-160.

Malik, R.P.S. (2000). Agriculture and water quality in India towards sustainable management. In: Water quality management and control of water pollution. pp. 73-85. Water Reports 21. FAO, Rome, Italy.

Mandrup-Poulsen, T. (1996). The role of interleukin-1 in the pathogenesis of IDDM. *Diabetologia* **39:** 1005-1029.

Markel, E., Nyakas, C. and Ormai, S. (1989). Nitrate induced changes in sensoro-motor development and learning behaviour in rats. *Acta Physiol. Hung.* **74:** 69-75.

Marshall, R.O., Disburger, H.J., MacVicar, R. and Hallmark, G.D. (1953). Studies on the effect of aeration on nitrate reduction by *Pseudomonas* species using N15. *J. Bacteriol.* **66:** 254-258.

Marshall, W. and Marshall, C.R. (1945). Action of nitrites on blood. *J. Biol. Chem.* **158:** 187-208.

Matt, G.E., Matt, G.E., Quintana, P.J., Hovell, M.F., Bernert, J.T. and Song, S. et al. (2004). Households contaminated by environmental tobacco smoke: Sources of infant exposures. *Tob Control* **13:** 29-37.

McCarty, P.L., Beck, L. and Amant, P.S. (1969). Biological denitrification of wastewater by addition of organic materials. Proc. 24th Industrial Waste Conference, Purdue University, 135, p. 1271.

Meiberg, J.B.M., Bruinenberg, P.M. and Harder, W. (1980). Effect of dissolved oxygen tension on the metabolism of methylated amines in hyphomicrobium X in the absence and presence of nitrate: evidence for aerobic denitrification. *J. Gen. Microbiol.* **120:** 453-463.

Michaud, D.S., Mysliwiec, P.A., Aldoori, W., Willett, W.C. and Giovannucci, E. (2004). Peptic ulcer disease and the risk of bladder cancer in a prospective study of male health professionals. *Cancer Epidemiol. Biomarkers Prev.* **13:** 250-254.

Mirvish, S.S. (1975). Formation of n-nitroso compounds: chemistry, kinetics, and *in vivo* occurrence. *Toxicol Appl. Pharmacol.* **31:** 325-351.

Mirvish, S.S. (1983). The etiology of gastric cancer: Intragastric nitrosamide formation and other theories. *J. Natl. Cancer Inst.* **71:** 631-647.

Mirvish, S.S. (1991). The significance of nitrate, nitrite, and n-nitroso compounds in human health. *In:* Nitrate Contamination: Exposure, Consequence, and Control, Bogardi, I., Kuzelka, R.D. and Ennenga, W.G. (Eds.). NATO ASI Series, Vol. 30. Springer Verlag. Berlin, pp. 253-266.

Mondal, N.C., Saxena, V.K. and Singh, V.S. (2008). Occurrence of elevated nitrate in the groundwaters of Krishna delta, India. *Afr. J. Environ. Sci. Technol.* **2:** 265-271.

Mueller, R.L., Hagel, H.J., Wild, H., Ruppin, H. and Domschke, W. (1986). Nitrate and nitrite in normal gastric juice. Precursors of the endogenous N-nitroso compound synthesis. *Oncology* **43:** 50-53.

Muhrer, M.E., Garner, G.B. and Pfander, W.H. (1959). The effect of nitrate on reproduction and lactation. *J. An. Sci.* **15:** 1291-1292.

Nagel, R.L. (2007). Methemoglobinemias and unstable hemoglobins. *In:* Cecil Text Book of Medicine, Goldman, L., Ausiello, D.A., Arend, W., Armitage, J.O., Clemmons, D., Drazen, J., Griggs, R., LaRusso, N., Newman, J. and Foster, E. (Eds.). 23th pp. 1226-1227.

NAS (National Academy of Sciences), Environmental Studies Board (1978). Nitrates: An Environmental Assessment. Washington, DC.

NAS (National Academy of Sciences), Safe Drinking Water Committee (1977). Drinking Water and Health. Washington, DC.

NEERI (1991). Water Mission Report on Participation of NEERI (National Environmental Engineering Research Institute), Nagpur, India.

Nitric Oxide, (1997). *In:* Smith, A.D., Datta, S.P. and Smith, G.H. (Eds.). Oxford dictionary of Biochemistry and Molecular Biology, Oxford University Press, New York, pp. 451.

Parslow, R.C., McKinney, P.A., Law, G.R., Staines, A., Williams, R. and Bodansky, H.J. (1997). Incidence of childhood diabetes mellitus in Yorkshire, northern England, is associated with nitrate in drinking water: an ecological analysis. *Diabetologia* **40:** 550-556.

PHED Habitat Survey, (1991-93). Nitrate concentration in water in the villages of Rajasthan. Rajasthan by Public health engineering department of Rajasthan.

Pomeranz, A., Korzets, Z., Vanunu, D., Krystal, H. and Wolach, B. (2000). Elevated salt and nitrate levels in drinking water cause an increase of blood pressure in schoolchildren. *Kidney Blood Press Res.* **23:** 400-403.

Porter, W.P., Jaeger, J.W. and Carlson, I.H. (1999). Endocrine, immune, and behavioral effects of aldicarb (carbamate), atrazine (triazine) and nitrate (fertilizer) mixtures at groundwater concentrations. *Toxicol. Ind. Health* **15:** 133-150.

Prakasa Rao, E.V.S. and Puttanna, K. (2006). Strategies for combating nitrate pollution. *Curr. Sci.* **91:** 1335-1339.

Prakasa Rao, E.V.S., Puttanna, K. and Singh, B. (2006). Nitrate levels in groundwater. Presented in Workshop on "Nitrogen in Environment, Industry and agriculture" March, 16-17, 2006. New Delhi, Indian National Science Academy and Society for Conservation of Nature.

Ptashekas, J.R. (1990). Ultrastructure of gastric epithelium after exposure to mild environmental agents (Sodium nitrate, Saprol). *J. Environ. Pathol. Toxicol. Oncol.* **10:** 297-300.

Revenga, C. and Mock, G. (2001). Dirty Water: Pollution Problems Persist. EarthTrends: The Environmental Information Portal, Source: Pilot Analysis of Global Ecosystems: Freshwater Systems.

RIVM, (1989). Integrated criteria document nitrate; effects. Appendix to RIVM Report No. 758473012. Bilthoven, Rijksintituut voor de Volksgenzondheid en Milieuhygiëne (National Institute of Public Health and Environmental Protection) (RIVM Report No. A758473012).

Robertson, L.A., van Niel, W.J., Torremans, R.A.M. and Kuenen, J.G. (1988). Simultaneous nitrification and denitrification in aerobic chemostat cultures of thiosphaera pantotropha. *Appl. Environ. Microbiol.* **54:** 2812-2818.

Robertson, R.M. and Robetson, D. (1996). Drugs used for the treatment of myocardial ischemia. *In:* Hardman, J.G., Limbirds, L.E., Molinoff, P.B. and Ruddon, R.W. (Eds.). Goodman & Gilman's The Pharmacological Basis of the Therapeutics. 9th edn. McGraw-Hill, New York, pp. 764.

Roediger, W.E., Deakin, E.J., Radcliffe, B.C. and Nance, S. (1986). Anion control of sodium absorption in the colon. *QJ. Exp. Physiol.* **71:** 195-204.

Royal Commission on Environmental Pollution (RCEP) 7th Report (1979). Agriculture and pollution, London, Her Majesty's Stationery Office, pp. 87-125, London.

Ruddell, W.S., Bone, E.S., Hill, M.J. and Walters, C.L. (1978). Pathogenesis of gastric cancer in pernicious anaemia. *Lancet* **1:** 521-523.

Samoli, E., Aga, E., Touloumi, G., Nisiotis, K., Forsberg, B., Lefranc, A., Pekkanen, J., Wojtyniak, B., Schindler, C., Niciu, E., Brunstein, R., Dodic Fikfak, M., Schwartz, J. and Katsouyanni, K. (2006). Short-term effects of nitrogen dioxide on mortality: an analysis within the APHEA project. *Eur. Respir. J.* **27:** 1129-1138.

Saul, R.L., Kabir, S.H., Cohen, Z., Bruce, W.R. and Archer, M.C. (1981). Reevaluation of nitrate and nitrite levels in the human intestine. *Cancer Res.* **41:** 2280-2283.

Schmitz, J.T. (1961). Methemoglobinemia—a cause of abortions? Preliminary report. *Obstet. Gynecol.* **17:** 413-415.

Schwiede, M., Duijnisveld, W.H.M. and Bottcher, J. (2005). Investigation of processes leading to nitrate enrichment in soils in the Kalahari Region, Botswana. *Physics and Chemistry of the Earth, Parts A/B/C* **30:** 712-716.

Shuval, H.I. and Gruener, N. (1972). Epidemiological and toxicological aspects of nitrates and nitrites in the environment. *Am. J. Public Health* **62:** 1045-1052.

Singh, B., Prakasa, Rao. E.V.S., Singh, Y. and Puttanna, K. (2007). Nitrate Pollution of Groundwater vis-à-vis Fertilizer Use in India. *In:* Agricultural Nitrogen Use and Its Environmental Implications. Abrol, Y.P., Raghuram, N. and Sachdev, M.S. (Eds.). I.K. International Publishing House Pvt. Ltd., New Delhi.

Singh, B., Singh, Y. and Sekhon, G.S. (1994). Fertilizer N use efficiency and nitrate pollution of groundwater in developing countries. 15th World Congress on Soil Science, 1994, Vol. 5a, pp. 174-191.

Sleiman, M., Gundel, L.A., Pankow, J.F., Jacob, P., Singer, B.C. and Destaillats, H. (2010). Formation of carcinogens indoors by surface-mediated reactions of nicotine with nitrous acid, leading to potential third-hand smoke hazards. *Proc. Natl. Acad. Sci.* **107:** 6576-6581.

Smith, A.D., Datta, S.P. and Smith, G.H. (1997). Nitric Oxide. *In:* Smith, A.D., Datta, S.P., Smith, G.H. et al. (Eds.). Oxford dictionary of Biochemistry and Molecular Biology, Oxford University Press, Oxford, New York-Tokyo, pp. 451.

Smith, R.A., Alexander, R.B. and Lanfear, K.J. (1994). Stream Water Quality in the Coterminous United States—Status and Trends of Selected Indicators During the 1980s. USGS Water Supply Paper 2400. Reston, Virginia: US Geological Survey.

Smith, R.L., Ceazan, M.L., Myron, N.D. and Brooks, H. (1994). Autotrophic, hydrogen-oxidizing, denitrifying bacteria in groundwater, potential agents for bioremediation of nitrate contamination. *Appl. Environ. Microbiol.* **60:** 1949-1955.

Soares, M.I.M., Belkin, S. and Abelivoch, A. (1988). Biological groundwater denitrification laboratory studies. *Water Sci. Technol.* **20:** 189-195.

Spalding, R.F. and Exner, M.E. (1993). Occurrence of nitrate in groundwater—a review. *J. Environ. Qual.* **22(3):** 392-402.

Sund, J.M., Wright, M.J. and Simon, J. (1957). Weeds containing nitrate cause abortion in cattle. *Agron. J.* **49:** 278-279.

Super, M., Heese, H.V., MacKenzie, D., Dempster, W.S. and du Plessis, J.J. (1981). An epidemiological study of well-water nitrates in a group of south west African/ Namibian infants. *Water Res.* **15:** 1265-1270.

Sutton, H.C., Roberts, P.B. and Winterbourn, C.C. (1976). The rate of reaction of superoxide radical ion with oxyhemoglobin and methemoglobin. *Biochem. J.* **155:** 503-510.

Sutton, P.M., Bridle, T.R. and Bedford, W.K. (1979). Nitrification and denitrification of an industrial wastewater. *JWPCF* **53:** 176-184.

Szökeová, E., Tajtáková, M., Mirossay, L., Mojzis, J., Langer, P., Marcinová, E., Petrovicová, J., Zemberová, E. and Bodnár, J. (2001). Effect of nitrates on active transport of iodine. *Vnitr. Lek.* **47:** 768-771.

Tajtáková, M., Semanová, Z., Tomková, Z., Szökeová, E., Majoros, J., Rádiková, Z., Seböková, E., Klimes, I. and Langer, P. (2006). Increased thyroid volume and frequency of thyroid disorders signs in schoolchildren from nitrate polluted area. *Chemosphere* **62:** 559-564.

Tannenbaum, S.R., Fett, D., Young, V.R., Land, P.D. and Bruce, W.R. (1978). Nitrite and nitrate are formed by endogenous synthesis in the human intestine. *Science* **200:** 1487-1489.

Terblanche, A.P.S. (1991). Health hazards of nitrate in drinking water and possible means of denitrification. *Water S.A.* **17:** 77-83.

Til, H.P., Falke, H.E., Kuper, C.F. and Willems, M.I. (1988). Evaluation of the oral toxicity of potassium nitrite in a 13-week drinking-water study in rats. *Food Chem. Toxicol.* **26:** 851-859.

Tonacchera, M., Pinchera, A., Dimida, A., Ferrarini, E., Agretti, P., Vitti, P., Santini, F., Crump, K. and Gibbs, J. (2004). Relative potencies and additivity of perchlorate, thiocyanate, nitrate, and iodide on the inhibition of radioactive iodide uptake by the human sodium iodide symporter. *Thyroid* **14:** 1012-1019.

Trachtman, H., Gauthier, B., Frank, R., Futterweit, S., Goldstein, A. and Tomczak, J. (1996). Increased urinary nitrite excretion in children with minimal change nephrotic syndrome. *J. Pediatr.* **128:** 173-176.

Turek, B., Hlavsova, D., Tucek, J., Waldman, J. and Cerna, J. (1980). The fate of nitrates and nitrites in the organism. *IARC Sci. Publ.* **31:** 625-632.

Ustyugova, I.V., Zeman, C., Dhanwada, K. and Beltz, L.A. (2002). Nitrates/nitrites alter human lymphocyte proliferation and cytokine production. *Arch. Environ. Contam. Toxicol.* **43:** 270-276.

Van Maanen, J.M., van Dijk, A., Mulder, K., de Baets, M.H., Menheere, P.C., van der Heide, D., Mertens, P.L. and Kleinjans, J.C. (1994). Consumption of drinking water with high nitrate levels causes hypertrophy of the thyroid. *Toxicol. Let.* **72:** 365-374.

Vander, A.J., Sherman, J.H. and Luciano, D.S. (1994). The digestion and absorption of food. Human Physiology. New York: pp. 561-600. McGraw-Hill, Inc, New York.

Verge, C.F., Howard, N.J., Irwig, L., Simpson, J.M., Mackerras, D. and Silink, M. (1994). Environmental factors in childhood IDDM. A population-based, case-control study. *Diabetes Care* **17:** 1381-1389.

Vermeer, I.T., Engels, L.G., Pachen, D.M., Dallinga, J.W., Kleinjans, J.C. and van Maanen, J.M. (2001). Intragastric volatile N-nitrosamines, nitrite, pH, and *Helicobacter pylori* during long-term treatment with omeprazole. *Gastroenterology* **121:** 517-525.

Violante, A., Cianetti, A. and Ordine, A. (1973). Adrenal cortex function during subacute poisoning with sodium nitrite. *Quad. Sclavo. Diagn.* **9:** 907-920.

Virtanen, S.M., Jaakkola, L., Räsänen, L., Ylönen, K., Aro, A., Lounamaa, R., Akerblom, H.K. and Tuomilehto, J. (1994). Nitrate and nitrite intake and the risk for type 1 diabetes in Finnish children. Childhood Diabetes in Finland Study Group. *Diabet. Med.* **11:** 656-662.

Vladeva, S., Gatseva, P. and Gopina, G. (2000). Comparative analysis of results from studies of goitre in children from Bulgarian villages with nitrate pollution of drinking water in 1995 and 1998. *Cent. Eur. J. Public Health* **8:** 179-181.

Wagner, D.A., Schultz, D.S., Deen, W.M., Young, V.R. and Tannenbaum, S.R. (1983a). Metabolic fate of an oral dose of 15N-labeled nitrate in humans: effect of diet supplementation with ascorbic acid. *Cancer Res.* **43:** 1921-1925.

Wagner, D.A., Young, V.R. and Tannenbaum, S.R. (1983b). Mammalian nitrate biosynthesis: incorporation of 15NH3 into nitrate is enhanced by endotoxin treatment. *Proc. Natl. Acad. Sci. USA* **80:** 4518-4521.

Wakida, F.T. and Lerner, D.N. (2005). Non-agricultural sorces of groundwater nitrate: review and a case study. *Water Res.* **39:** 3-16.

Waldman, S.A. and Murad, F. (1987). Cyclic GMP synthesis and function. *Pharmacol. Rev.* **39:** 163-196.

Walker, R. (1995). The conversion of nitrate into nitrite in several animal species and man. In: Health aspects of nitrate and its metabolites (particularly nitrite). Proceedings of an international workshop, Bilthoven (Netherlands), 8-10 November 1994. Strasbourg, Council of Europe Press, Strasbourg, pp. 115-123.

Water quality standard (1991). Indian Standard Drinking Water–Specification IS 10500.

Weisenburger, D.D. (1990). Environmental epidimiology of Non-Hodgkin's Lymphoma in eastern Nebraska. *Am. J. Ind. Med.* **18:** 303-305.

Weyer, P.J., Cerhan, J.R., Kross, B.C., Hallberg, G.R., Kantamneni, J., Breuer, G., Jones, M.P., Zheng, W. and Lynch, C.F. (2001). Municipal drinking water nitrate level and cancer risk in older women: the Iowa Women's Health Study. *Epidemiology* **12:** 327-338.

WHO (1977a). Nitrate, Nitrite and N-Nitroso compounds. In Environmental Health Criteria 5, Geneva, p. 15.

WHO (1977b). Nitrate, Nitrite and N-Nitroso compounds. In Environmental Health Criteria 5. Geneva, p. 82.

WHO (1977c). Nitrate, Nitrite and N-Nitroso compounds. In Environmental Health Criteria 5. Geneva, p. 73.

WHO (1977d). Nitrate, Nitrite and N-Nitroso compounds. In Environmental Health Criteria 5, Geneva, p. 78.

WHO (1977e). Nitrate, Nitrite and N-Nitroso compounds. In Environmental Health Criteria 5, Geneva, p. 37.

WHO (1985a). Health hazards from nitrate in drinking-water. In Report on a WHO meeting, Copenhagen, 5-9 March 1984. Copenhagen, WHO Regional Office for Europe (Environmental Health Series No. 1), Copenhagen, pp. 14-16.

WHO (1985b). Health hazards from nitrate in drinking-water. In Report on a WHO meeting, Copenhagen, 5-9 March 1984. Copenhagen, WHO Regional Office for Europe (Environmental Health Series No. 1), Copenhagen, pp. 23-40.

WHO (1985c). Health hazards from nitrate in drinking-water. In Report on a WHO meeting, Copenhagen, 5-9 March 1984. Copenhagen, WHO Regional Office for Europe (Environmental Health Series No. 1), Copenhagen, pp. 19.

WHO (1996). Toxicological evaluation of certain food additives and contaminants. Prepared by the Forty-Fourth Meeting of the Joint FAO/WHO Expert Committee on Food Additives (JECFA). Geneva, World Health Organization, International Programme on Chemical Safety (WHO Food Additives Series 35).

WHO (2003). Nitrate and nitrite in drinking water. Background document for development of WHO guidelines for drinking water Quality. World Health Organization, Geneva.WHO/SDE/WSH/04-03/56. English.

WHO (2004). Chemical aspects. In Guidelines for drinking-water quality. *Geneva* **1:** 145-196.

Wilson, G.L., Mossman, B.T. and Craighead, J.E. (1983). Use of pancreatic beta cells in culture to identify diabetogenic N-nitroso compounds. *In Vitro* **19:** 25-30.

Winterbourn, C.C., McGrath, B.M. and Carrell, R.W. (1976). Reaction involving superoxide and normal and unstable haemoglobins. *Biochem. J.* **155:** 493-502.

Zaldivar, R. (1977) .Nitrate fertilizers as environmental pollutants; positive correlation between nitrates ($NaNO_3$ and KNO_3) used per unit area and stomach cancer mortality rates. *Experientia* **33:** 264-265.

4

Nitrate Content in the Edible Parts of Vegetables: Origin, Safety, Toxicity Limits and the Prevalence of Cancer in Iran

M.J. MALAKOUTI[1], SH. LADAN[1] AND S.J. TABATABAEE[2]

[1]*Soil Science Department, Tarbiat Modares University, Tehran, Iran*
[2]*Horticultural Science Department, Tabriz University, Tabriz, Iran*
e-mail: mjmalakouti@modares.ac.ir; mjmalakouti@hotmail.com

ABSTRACT

The mineral nutrient needed in the greatest abundance by plants is nitrogen (N), which is mainly absorbed in the form of nitrate (NO_3). The higher content of NO_3 in the edible parts of vegetables, which are a major source of human diet, is believed to be potentially hazardous to human health. The role of vegetables in human diet is very important in Iran because daily use of vegetables, regardless of other NO_3 sources such as drinking water and processed foods, is about 782 g day^{-1}. Imbalanced fertilization accelerates NO_3 accumulation in the edible parts of agricultural products. The NO_3 increase in the soil caused by either the overuse of N fertilization or the nitrification process in the soil stimulates NO_3 uptake by the vegetable crops. It results in the accumulation of large amounts of NO_3 in the edible parts of the vegetables. An average acceptable daily intake of NO_3 is 3.7 mg kg^{-1} b. w. day^{-1}, equivalent to 222 mg NO_3 per day for an adult of 60 kg weight. It shows that the daily NO_3 intake depends on the dietary exposure. As long as the major part of people in Iran consumes high amounts of leafy vegetables, it can lead to the daily NO_3 intake exceeding the acceptable limit. Many factors affect NO_3 content in the plants, of which the most important ones are imbalanced fertilization (both the forms and rates of N fertilizer), cultivar, yield, NO_3 concentration in irrigation water, soil organic matter content, temperature, light intensity and time of harvesting, i.e., both the season and the time of day (morning/afternoon). In order to determine the allowable limit of NO_3 in the edible parts of vegetables such as radish, parsley, coriander, lettuce, basil, leek, cress, cucumber, melon, tomato, onion, green onion and potato all around the year, i.e. winter, autumn, summer and spring, numerous experiments were carried out all over the country during the past two decades. The results revealed that potato, onion and tomato had the lowest NO_3 in contrast to lettuce, spinach, celery, radish and cress. Lettuce and radish had higher potential to accumulate NO_3, so that the NO_3 content in these latter vegetables was higher than the acceptable daily intake (ADI) limit for a person of an average 60 kg weight (up to 7.0 mg kg^{-1} b. w. day^{-1} of body weight), if consumed more than 100 g day^{-1}. Harvesting periods had no effect on the NO_3 content of both potato and onion. However, NO_3 content varied considerably in different geographical areas, but was mostly lower than permitted level (60 mg kg^{-1} FW), especially in those farms in which balanced fertilization was practiced. The maximum permitted limit of NO_3 levels suggested for leek, parsley, coriander and basil were 200, 500, 600 and 800 mg kg^{-1} FW, respectively. Considering the fact that vegetables are consumed in high amounts in Iran, the NO_3 content of vegetables should be monitored carefully as the higher content of NO_3 is fatal to human health, such as by producing nitrosamine compounds, which have been known to be powerful carcinogens. According to the latest findings in northern region of Iran, it has one of the highest rates of esophageal cancer (EC) in the world. At present 70,000 people are suffering from cancer, mainly gastroesophageal cancer. Furthermore, reconsidering N fertilizers subsidy, balanced fertilization, legislation on sampling, analysis and the maximum permitted level of NO_3 needs to govern allowable NO_3 content in the vegetables. We humbly request the authorities of the Ministry of Jehad-e-Agriculture to improve the status of plant nutrition and enhance balanced fertilization to meet the requirements of food security within the context of safer agricultural products. We also request physicians and country's health authorities to give serious considerations, firstly to NO_3 as the number one risk factor on gastroesophageal cancer and also focusing on pesticide residues as well as cadmium (Cd) contaminants as the main sources for this type of cancer and other diseases. From agricultural point of view, we strongly believe that there is a positive link between gastroesophageal cancer and unsafe agricultural products processed foods and drinks which contain amounts beyond and above allowable limits of contaminants, particularly NO_3, pesticide residues and Cd, in descending order.

Keywords: Accumulation of NO_3, Imbalanced fertilization, Cultivar, N-fertilizers, Urea, Edible parts of vegetables, Gastroesophageal cancer, NO_3 permitted level and human health

INTRODUCTION

The main concern in the 3rd millennium in agricultural production should be the determination of production capacity of soil resources. It should crucially monitor the changes in the soil resources quality under intensive farming for sustained agricultural production, increasing fertilizer use efficiency, increasing total production of agricultural crops, environmental issues and enrichment of agricultural products to improve food quality. Efforts to improve the yield and quality of agricultural products are deemed to be increasingly necessary because of the growth in population and increasing demand for more and better foodstuffs. Soil quality plays a key role in achieving food security and agricultural sustainability. Soil health and sustainable agriculture will determine the quality of the environment and consequently the health of plants, animals and humans. For sustainable agriculture, it is not wise to tolerate hidden deficiency symptoms or imbalanced fertilization because in such situations a great yield loss, i.e., at least 30%, would be expected. One of the main principles of sustainable agriculture is to return to the soil those nutrients that have been removed from it by crop (Malakouti and Mashayekhi 1996; Malakouti 2008; Malakouti et al. 2008, Malakouti et al. 2010; Malakouti 2011).

In the third millennium, due to environmental and economic concerns, the challenge of increasing yield, nitrogen use efficiency (NUE) and higher crop quality with lower amounts of contaminants, especially nitrate (NO_3), was addressed and taken into consideration. Since the farmer's conventional fertilization practice is mostly the application of N and P-fertilizers without using soil testing analysis results, the use of urea which constitutes 80% of the 2.70 million tons of annual N-fertilizers use, is done with no consideration given to crop quality because of heavily subsidized N-fertilizers prices (Malakouti 1996; Malakouti et al. 2008).

Among the essential nutrients, nitrogen (N) is the most limiting factor (bottleneck) for plant growth. Most plants need N in large amounts as an essential component of proteins, nucleic acids and other cellular constituents. However, excessive use of N-fertilizers has led to the accumulation of NO_3 in the edible parts of agricultural production, especially leafy vegetables (Malakouti 1996, Malakouti et al. 2004). Nitrate is added by waste discharge, animal slurries and N-fertilizers. Nitrate in the soil is soluble and readily migrates into ground and surface water (Environmental Protection Agency 2010).

The attention shifted to sustainable agriculture (agriculture with the balanced application of nutrients) since the middle of 80s, as a result of demands for higher quality, safe foods, economic pressures and the intensity of environment quality. The rational use of different kinds of nutrients especially fertilizers is one of the ways along the policies of sustainable development of agriculture, which is absolutely important nowadays. During the last two decades, some fundamental changes have been made in the structure of crop production, and these changes have been effective due to the balanced application of chemical fertilizers, and pesticides, the execution of managements and scientific methods in agricultural activities.

According to the Bureau of Statistics and Information Technology Office annual report (2007), Iranian farmers produced almost 21 million tons of vegetables, i.e., potato (4.83), onion (1.68), tomato (4.78), leafy vegetables (2.49), melon (1.58), water melon (3.26), cucumber (1.72). By assuming one million ton export, 70 million Iranian used almost 20 million tons of all types of vegetables. So, the average daily vegetable consumption will be around 782 g day^{-1}. Obviously, these figures show the importance of vegetables in our daily diet. Because of the function of N in yield enhancement, relative ease of the production of N-fertilizers and N's mobility, the use of N-fertilizers is high, (88 million tons per year (Heffer, 2008). Due to this fact, a great deal of NO_3 is leached from soil and finally it leads to groundwater pollution. The large amount of NO_3 in animals leads to abortion and decrease in milk production. In addition, it affects the growth in humans, especially newborn babies.

Due to the overuse of N-fertilizers, NO_3 accumulation in plants especially in the edible part of vegetables, is observed that results in producing some harmful compounds such as nitrosamine (Fig. 4.1). This compound combines with hemoglobin and causes oxygen shortage in the blood of humans. Large amount of nitrate is not harmful for plants, but in humans it leads to various diseases. The disease results from using vegetables with higher NO_3 contents and contaminated drinking water, which contains excessive amount of NO_3. Human is safe from the harms of NO_3 just in case of the condition in which he uses lower amounts of NO_3 in his daily food regime (Malakouti 1996; Anac and Martin-Preve 1999; Malakouti et al. 2004; Malakouti et al. 2008).

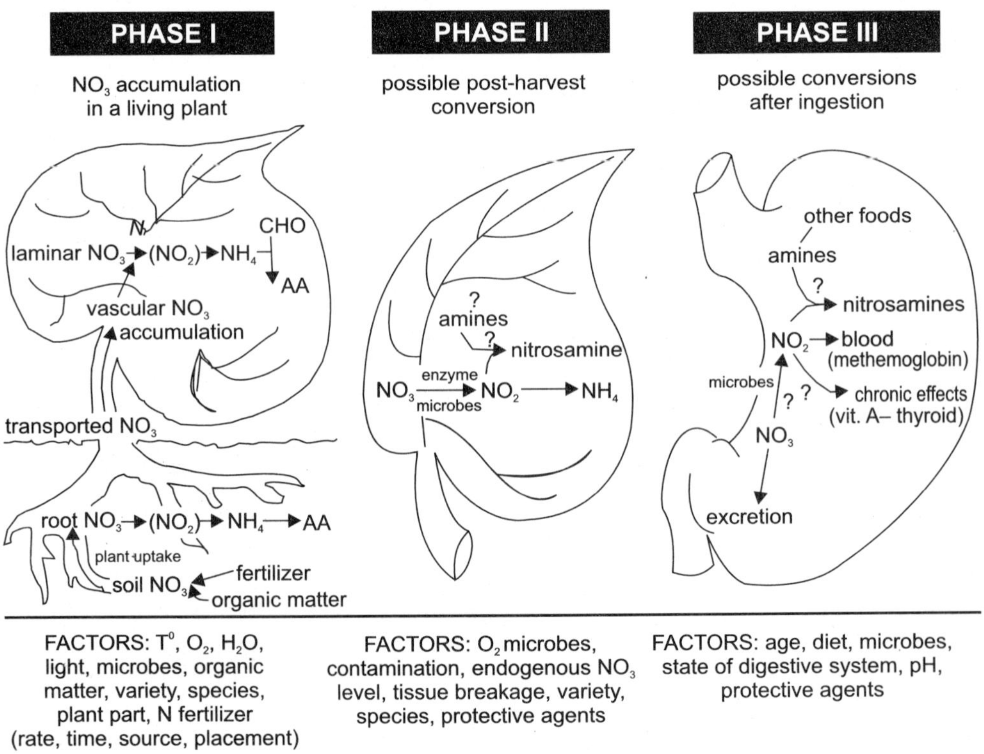

Fig. 4.1. Possible post-harvest conversion of NO_3 in the leaves and human stomach (Anac and Martin-Preve 1999).

In this paper, factors affecting NO_3 accumulation in the plants will be discussed However, growers and appropriate standards can play a key role in reducing NO_3 in the plants.

SOURCES OF NITRATE

Nitrate is an inorganic chemical that is highly soluble in water. Most N containing materials tend to be converted to NO_3. Nitrates also occur naturally in the environment, mineral deposits, soil, seawater, freshwater systems, and the atmosphere. Nitrate and nitrite (NO_2) are commonly used as a preservative and for color enhancement of processed meats, although the amounts added to these products have been substantially reduced from the levels once used. Nitrate itself, in its initial form, is non-toxic, but it becomes toxic by conversion to NO_2 in the digestive system. Food is usually the major source of NO_3 exposure. Nitrate intake from a typical US diet provides an average of 75 to 100 mg day^{-1} of NO_3. Vegetables, particularly spinach, celery, beets, lettuce, and root vegetables are responsible for most of the dietary intake. Nitrate is a health hazard because of its conversion to NO_2. Once ingested, conversion of NO_3 to NO_2 takes place in the digestive system in all age groups, and in the gastrointestinal tract of infants. Infants convert approximately double, or 10% of ingested NO_3 to NO_2 compared to 5 percent conversion in adults. No cases of methemoglobinemia have been reported when water contained less than 10 mg kg^{-1} of NO_3. The majority of cases involve exposure to levels in drinking water exceeding 50 mg kg^{-1}. It is believed that after NO_3 is converted to nitrite in the body, it can react with certain amine containing substances found in food to form nitrosamines, which are known to be potent cancer causing chemicals. Nitrosamine formation is inhibited by antioxidants that may be present in food such as vitamin C and vitamin E. In some laboratory studies in which rodents were given high levels of nitrites along with amine-containing chemicals, cancers of the lung, liver, and esophagus were observed. However, cancer was not observed in animals given NO_3. The presence of NO_3 in food and drinking water is considered a major public health problem. Two main types of public health risk due to NO_3 exposure are distinguished acute (direct) effects, i.e., methemoglobinemia and chronic (long-term) effects, i.e., gastrointestinal cancers. It should be noted again that the toxicity of NO_3 is in fact caused primarily by NO_2 formed in the body. Nitrate itself is practically non-toxic apart from the disruption of the osmotic and electrolytic balance in stomach and intestines inherently caused by salts. The NO_2 formed can oxidize hemoglobin to methemoglobin resulting in methemoglobinemia (Office of Health and Environmental Assessment, 1990). But according to Laegreid et al. (1999), NO_2 is one of the substances that transforms hemoglobin (the oxygen carrier in the blood) into inactive methemoglobin, and there are some reports that this type of disease has been reduced in the past decades.

There is no compound like NO_3 that makes it a controversial issue. Therefore, NO_3 is considered not just a chemical fertilizer, but also a threat to human health and environment. About 75% of the N absorbed by plants is in the form of NO_3 and this ion plays an important role in the metabolisms of plants. When ammonium-based fertilizers like urea are applied to the soil, soil microorganisms convert it to NO_3. The rate of these changes is a function of soil properties. When ammonium fertilizers such as ammonium sulphate are added to the soil, they are oxidized by the microorganisms in two steps:

$$NH_4 + 3/2\ O_2 \longrightarrow NO_2 + H_2O + 66KCal$$

$$NO_2 + 1/2\ O_2 \longrightarrow NO_3 + 17KCal$$

A part of soil NO_3 is absorbed and fixed by the microorganisms from atmosphere directly, and another part of soil NO_3 content is the result of the mineralization of soil organic matter. Out of the total soil NO_3 content, a part is absorbed by the plant and enters its metabolism, another part is leached through irrigation and rainfall that eventually reaches groundwater, and the last portion through denitrification is transformed into N gas and enters the atmosphere (Malakouti et al. 2008).

FUNCTION OF NITRATE IN PLANTS

Uptake of NO_3 by the roots is regulated by transpiration (mass flow), soil moisture and soil temperature (diffusion). Nitrate reductase (NR) in the leaves catalyzes the reduction of NO_3 to amino acids, which are subsequently converted to N-sugars and chlorophyll components (Marschner, 1995). NO_3 reduction is accelerated by the conditions that promote photosynthesis—favorable temperatures, high CO_2 concentrations and high light levels. Surplus NO_3 is transiently stored in the vacuole with leaf NO_3 levels peaking at night. Approximately 33% of the leaf N is in the Calvin cycle enzymes, while 25% is in chlorophyll and related compounds (Kumar et al. 2002). Increased NO_3 uptake and leaf N contents enhance photosynthetic activity by increasing the number of mesophyll chloroplasts. NR activity is induced by high levels of NO_3 in the leaves. Nitrogen is mostly (75%) absorbed in the form of NO_3 by plants. Nitrate is assimilated by the plants to synthesize amino acids and protein as follows (Marschner, 1995; Malakouti et al., 2008):

$$NH_3 \rightarrow \text{carbon skeleton} \longrightarrow \text{glutamic acid} \longrightarrow \text{amino acids} \longrightarrow \text{proteins}$$

The proteins have many functions in the metabolism of plants. In addition to contributing in proteins' structure, the existing N in a plant forms a part of chlorophyll. The lack of NO_3 sources in soils and limiting factors for absorbing this ion cause the deficiency symptoms to appear in the old leaves. High mobility of N in plants makes the young leaves remain green and the old leaves become yellow under deficiency condition. When roots are not able to absorb sufficient amounts of N, nitrogen compounds (proteins) break down in the old tissues, and redistribute to young leaves (Malakouti et al. 2008). This process causes the senescence of old leaves, reducing rate photosynthesis.

NITRATE ACCUMULATION IN PLANTS

Most of the plants such as leafy vegetables, fruit bearing vegetables and tuberous ones can easily accumulate NO_3 in their leaves, roots and stems. The rate of accumulation depends on the yield, plant species, form, rates and time of N-fertilizers application, plant's age and environmental conditions. When NO_3 content is low, the translocation of NO_3 from roots to the aerial parts seems to be low. By increasing the NO_3 content, transfers to stems and leaves is stimulated. Therefore, under the condition of applying large amounts of N-fertilizers, the capacity of NO_3 reduction decreases in roots and stems and finally excess amounts of NO_3 transfers to leaves. According to the fact that leaves act as the source, during the production stage, the production of hydrocarbon compounds, which is necessary for NO_3 reduction, decreases. As a result, the capacity of NO_3 in leaves lowers during the production stage. Due to the high mobility of NO_3, especially at the time of applying a great deal of N-fertilizers (in the form of NO_3), plant does not have the ability to reduce the excess NO_3. In fruit growth stages, they act as main sink and consume the compounds which are produced in leaves. Moreover, the carbon in the fruit is mostly consumed in respiration process and fruit cannot make use of these hydrocarbon compounds in order to reduce NO_3. In general, there is a limited capacity for NO_3 reduction in fruits like other parts of a plant. Furthermore, in mature leaves, the NO_3 reductase activity is too low, which ends up in decreasing NO_3 content in leaves (Malakouti et al. 2005).

Nitrogen fertilizers, after use in the soil, are finally changed into NO_3 by some nitrifying bacteria such as *nitrosomonas* and *nitrobacter*. In normal conditions, plants after NO_3 absorption, reduce it in the root and transform it into ammonium compounds as well. Then, these compounds can be transferred to different parts of the plant by the use of vessel system. As a result of higher consumption of N fertilizers, the NO_3 in the soil increases. Plant do not have the required ability to reduce this substance, and this combination through mass flow and vessel system will enter the plant tissues. In spinach, as will be discussed later, most of this NO_3 accumulates in its short stem and petiole and only a little amount of it enters the processes such as synthesis and decomposition of proteins. During the day, the activity of NO_3 reductase as well as photosynthesis and protein synthesis increases by the use of light; in both cases the accumulated NO_3 in tissues will be used and decreased. Therefore, in the crops which are harvested in the afternoon, the accumulation of NO_3 is less (Malakouti 2002). Nitrate accumulation in plants is a natural phenomenon and it happens when NO_3 accumulation in plants is more than its absorption and decomposition. The capacity of NO_3 accumulation by plants is regulated by plant's inheritance ability and it changes because of environmental factors, management, fertilization and cultivation activities. Nitrate accumulation in food is an interesting topic for the experts in the field of human and animal nutrition. The transformation of NO_3 into nitrite in the body leads to nitrite toxicity, especially in ruminants and newborn babies. The clearest symptom of this toxicity is methemoglobinemia disease in which hemoglobin changes into methemoglobin. In this process Fe^{+2} changes to Fe^{+3} which causes disorder in the transfer of oxygen in body and suffocation happens especially in newborn babies, resulting in death. Furthermore, because of the continuous consumption of vegetables and water containing a lot of NO_3, with amino compounds in the stomach react with NO_3 producing nitrosamine, which is a poisonous substance that can cause cancer (Malakouti 2002). One of the ways of NO_3 accumulation in crops is irregular use of N fertilizers. In general, vegetables (radish, spinach, celery, beet root, lettuce, carrot and cauliflower), corn and alfalfa are the products in which NO_3 accumulation is more. In contrast, the possibility of NO_3 accumulation in edible parts of fruits, onion and potato is very little and usually it does not create any problem. But in conserved products, especially beans, tomato and the fruits which produce acid, NO_3 is used as an oxidized factor, and in this condition NO_3 content can reach toxic level. In leafy vegetables like lettuce, spinach and cabbage, NO_3 seems to be lower up to 50% when they are harvested in late afternoon. This is due to the reduction of NO_3 caused by light dependent metabolism of NO_3 during daytime (Malakouti 2002).

A number of significant factors have been shown to affect NO_3 accumulation in plants.

Nitrogen Fertilizers

The amount, form, time and method of N fertilizers application influence NO_3 accumulation. A straight relationship between N fertilizer and NO_3 accumulation has been observed for many plant species particularly under imbalanced fertilization. In fact, N fertilizers are the most important factors causing accumulation of nitrate in vegetables and other plants. Studies have revealed that the existence of excess N causes NO_3 accumulation in the edible parts of vegetables (Malakouti 1996; Malakouti 2000; Malakouti 2002; Malakouti et al. 2005; Malakouti et al. 2007; Malakouti et al. 2008). Andersen and Nielsen (1992) introduced a new cultivation method for the production of vegetables with low NO_3 content. In this new technique, N supply was reduced in proportion to the daily uptake of vegetables. The results showed that despite 30–40% decrease of NO_3 in lettuce, the yield and quality of lettuce remained unchanged. Tabatabaei (1998) studied the effects of balanced fertilization on NO_3 reduction in potato and found that (a) there is a positive relationship between the amounts of N-fertilizer and accumulation of NO_3 in potato tubers and (b) there is a significant difference in NO_3 among potato cultivars, and by balanced and on-time fertilization, crop quantity and quality increase, and accumulation of NO_3 decreases below the allowable permitted level (Figs. 4.2 and 4.3). In Figs. 4.3 and 4.4 imbalanced and balanced fertilization effects can be seen on the quality of potatoes.

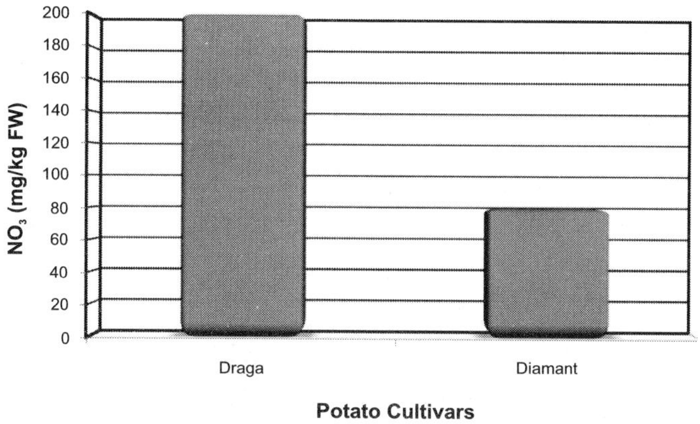

Fig. 4.2. Nitrate content of two cultivars of potato supplied with 200 kg N/ha.

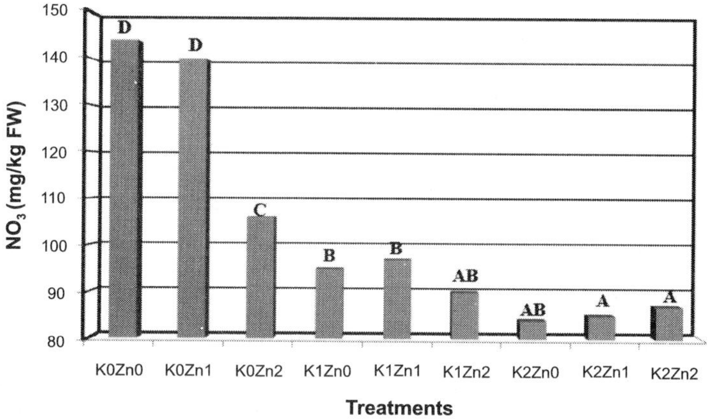

Fig. 4.3. Effects of balanced fertilization on the reduction of NO_3.

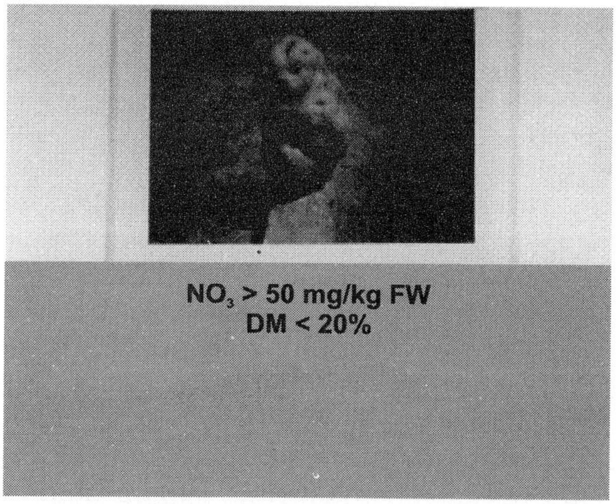

Fig. 4.4. Effects of overuse of N fertilization on nitrate content of potato and reducing DM percentage.

Fig. 4.5. Effects of balanced fertilization on the reduction of NO_3 and increasing DM percentage in potato tubers.

In order to investigate the effect of N fertilizers on NO_3 accumulation in vegetables, in the past 2 decades, many field experiments were carried out all over Iran. The results revealed that: (a) N fertilizer types and rates, imbalanced fertilization had the most significant effects on NO_3 accumulation in all types of vegetables; (b) in leafy vegetables such as lettuce and spinach, the most NO_3 content was observed in petioles and stem, and the least NO_3 content was observed in the leaf blade, outer leaves and root; (c) in both vegetables, morning samples contained much higher NO_3 content in comparison with afternoon's samples; (d) cultivars accumulated various levels of NO_3; (e) accumulation of NO_3 in leafy vegetables was maximum; and (f) vegetables grown in the spring had lower NO_3 accumulation in their edible parts (Malakouti 2002).

Bakhtiyari et al. (2004) carried out an experiment in order to analyze the yield and compare NO_3 content between the cucumbers grown hydroponically and conventionally. The results revealed that while the mean NO_3 content for cucumbers produced in hydroponics system equaled to 43 mg kg^{-1} FW, it increased up to 208 mg kg^{-1} FW in the field produced cucumbers. In general, cucumbers grown hydroponically had a lower NO_3 content and higher dry matter percentage and were of much better quality. Rajabzadeh (2004), in order to determine the maximum allowable NO_3 content in vegetables in Tehran's local markets, carried out an experiment. Samples from 10 types of vegetables (radish, onion, green onion, parsley, coriander, lettuce, leek, cress, potato and basil) were taken in winter, autumn, summer and spring. Samples were gathered every two weeks from Tehran Markets and their NO_3 contents were measured by the use of Diazo method. An experiment was carried out to determine the maximum permissible content of NO_3 in vegetables. Results have shown that potato, onion and tomato had the least NO_3 accumulation in contrast to lettuce, spinach, celery, radish and cress which accumulated maximum NO_3. In winter cultivation, their NO_3 content was higher due to daytime shortness. The results revealed that in cress samples, there was no meaningful difference between the NO_3 content in different periods of time and NO_3 content was measured up to 3500 mg kg^{-1} on a fresh weight basis, which was higher than the permitted level (500-700 mg kg^{-1} FW). Lettuce had higher NO_3 accumulation, for which the maximum permitted content of NO_3 is suggested at 500-700 mg kg^{-1} FW. In lettuce, the major part of NO_3 accumulated in the outer and old leaves. In potato there was no meaningful difference between NO_3 contents in different harvesting periods, but NO_3 contents varied considerably in the farms and different areas and the maximum permissible content is suggested at 50 mg kg^{-1} FW. The maximum permitted limit of NO_3 suggested for leek, parsley, coriander, and basil was 200, 500, 600 and 800 mg kg^{-1} FW, respectively. In radish, the maximum permitted content of NO_3 is considered at the amount of 1000 to 1200 mg kg^{-1} FW. The main reason for higher NO_3 in these vegetables is the use of excess amounts of N fertilizers. In tomato, potato, garlic and onion, there was no noticeable difference in their NO_3 contents among different periods of harvesting, and their maximum permissible levels were suggested to be 40, 50 and 60 mg kg^{-1} FW, respectively.

In order to determine the permitted limit of NO_3 in the edible parts of vegetables, too many experiments were carried out all over the country during the past two decades. The results revealed that in leafy vegetables such as spinach, lettuce, celery and radish, accumulation of NO_3 was higher than the acceptable daily intake (ADI) limit for an average 60 kg person if consumed more than 100 g day^{-1}. In beet root, a study showed that more than 56 kg ha^{-1} causes NO_3 accumulation. Other researchers have found that NO_3 accumulation happens rarely with less than 100 kg ha^{-1}N. Some researchers have proved that if 90 kg ha^{-1} N or less is used, NO_3 content seldom reach higher levels than its toxic levels. Application of slow-released fertilizers such as sulfur coated urea (SCU), which release their N at a low rate, can decrease NO_3 accumulation. The use of phosphorus (P) has a little effect of NO_3 accumulation, whereas the use of potassium (K) and zinc (Zn), due to their positive roles in increasing N recovery and plant enzymes activities, usually results in more biomass production and less NO_3 accumulation (Malakouti et al. 2005; Malakouti and Bybordi 2006).

Mahmoudi (2005) studied the effect of amount and sources of N fertilizers on NO_3 accumulation and yield of lettuce. In this experiment, the treatments included four N levels (0, 100, 200, and 300 kg ha^{-1} N) with four fertilizers sources (urea, ammonium nitrate, SCU and urea). The results demonstrated the significant effect of treatments containing different amounts of N on the yield, dry matter percentage, protein, chlorophyll index and the content of N, NO_3, P, K, Zn and Mn. The results also revealed that the treatments from different N sources had a meaningful effect on the yield, NO_3 content and chlorophyll index as well. Maximum yield was obtained by the application of 200 kg ha^{-1} N from the ammonium nitrate source. Nitrogen increase from N_{200} to N_{300} has shown no significant increase in lettuce yield. Maximum amount of NO_3 accumulation in leaves was observed by the application of 300 kg ha^{-1} N from the ammonium nitrate source. Sulfur coated urea as N source led to least NO_3 accumulation in lettuce leaves. The results from root samples have shown a meaningful effect of treatment with different amounts of N on protein, dry matter, NO_3 and P contents. The overall results demonstrated that an application of 200 kg ha^{-1} SCU would be favorable to obtain an acceptable yield as well as lower the NO_3 content and also for other properties such as protein, chlorophyll index and nutrient content.

Pishnamaz (2005) studied the ways to increase nitrogen use efficiency (NUE) and minimize NO_3 accumulation in high yielding potato in Zanjan province. The results revealed that all N fertilizer treatments had significant differences at 1% level. While the average NO_3 in the control plots was 22 mg kg^{-1} FW, it increased up to 94 mg kg^{-1} FW in the urea and ammonium nitrate plots, but NO_3 content decreased significantly (40 mg kg^{-1} FW) with pre-plant SCU and top-dressing of urea before potato flowering stage. In another experiment, the effect of different levels of nutrients in both soil and hydroponics cultures with two cucumber varieties was studied and it was found that the weights of cucumber fruits in soil treatment was higher than hydroponics, 14 and 12%, respectively. In hydroponics treatment, the number of leaves and length of fruit were higher than the soil medium; in addition, the content of NO_3 did not change significantly in different nutrient solutions. While the average NO_3 level in hydroponics was 33 mg kg^{-1}on fresh weight basis, it increased up to 43 mg kg^{-1}. In other words, the content of cucumber's NO_3 in soil medium was 23% higher than hydroponics media. According to his findings, the content of NO_3 in cucumber was lower in hydroponics culture than soil cultivation (Malakouti et al. 2008).

To evaluate the effect of Zn and K fertilizer sources and rates on the yield and quality, especially percentage of dry matter (DM) and NO_3 level in potato and onion, three different field experiments were carried out in 2001 to 2003 in East Azarbaijan and Zanjan Agricultural Research Centers. The treatments included: T_1= Farmers' conventional fertilizer use method (NP); T_2= NP+ potassium sulfate at the rates based on soil test and T_3= T_2+ zinc sulfate. In the first experiment, the analysis of variances showed significant differences at 1% probability level in the potato yield, the size of tubers, the starch content, and K and Zn levels in the tubers. While in the control plots, the content of NO_3 was 250 mg kg^{-1} FW, this value decreased to 50 mg kg^{-1} FW in the balanced fertilization plots. There were significant differences at 5% probability level in the NO_3 content of the tubers as well as DM with different treatments. The lowest level of NO_3 and the highest yield and DM are shown in Figs. 4.6 and 4.7. While in the control plot, the content of NO_3 was 320 mg kg^{-1} FW, this value decreased to 85 mg kg^{-1} FW. The results of the third experiment revealed that the yield, TSS, ascorbic acid and NO_3 content in onion bulbs varied significantly. The highest yield was obtained with 50 kg ha^{-1} zinc sulfate. As for quality, soil application of 50 kg ha^{-1} zinc sulfate resulted in onions with higher DM and higher contents of ascorbic acid (14 mg kg^{-1}) and titratable acidity. The highest NO_3 content (220 mg kg^{-1} FW) in the onion bulbs was obtained in the control plots, while the lowest NO_3 content (80 mg kg^{-1} FW) was obtained with 50 kg ha^{-1} zinc sulfate application (Malakouti and Bybordi 2006). Hashemimajd and Fathi (2006) estimated the amounts of nitrate daily intake by the consumers in Ardabil province. They selected 50 families (240 persons) and measured their daily NO_3 intake from different food sources and found that about 6.16 mg per kg body weight (allowable level is around 3.7 mg kg^{-1}NO_3 g day^{-1}) and more than 60% of NO_3 were from the vegetables.

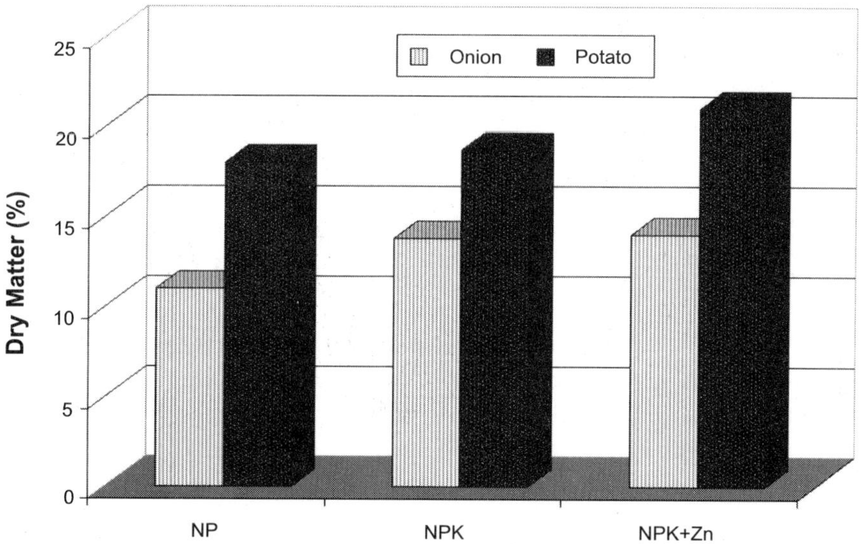

Fig. 4.6. The role of balanced fertilization on increasing the percentage of dry matter.

Anjana et al. (2006) studied accumulation of NO_3 in plants and concluded that: (a) under excess application of N fertilizers, vegetables accumulate high level of NO_3, even higher than ADI (b) main factors responsible for NO_3 accumulation in vegetables are mainly nutritional, environmental and physiological. Among these factors, N fertilizers and light intensity have been identified as the major factors that influence the NO_3 content in vegetables and (c) 77 to 88% of the total NO_3 daily intake comes from vegetables and fruits. Their findings mostly corresponded with ours, but they did not mention about the importance of balanced fertilization in reducing NO_3 in the edible parts of vegetables.

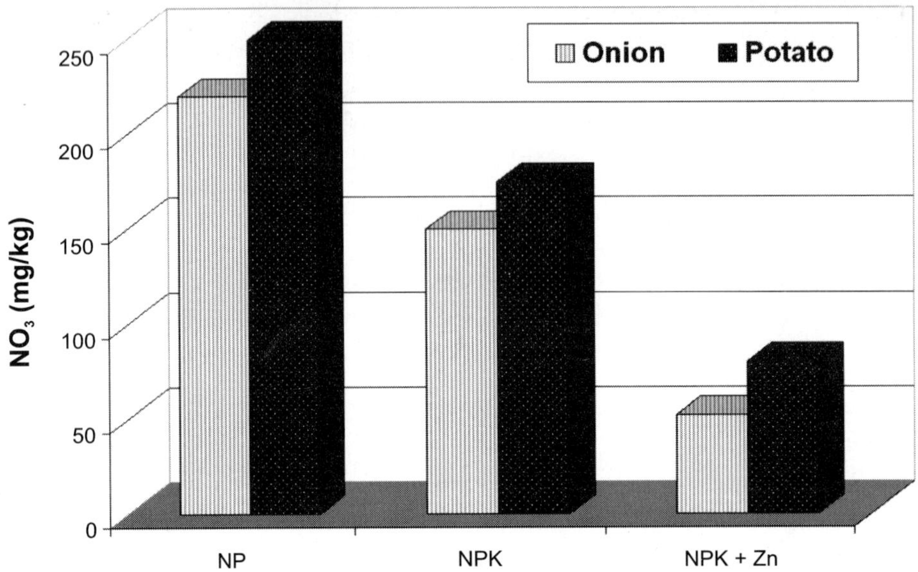

Fig. 4.7. Effect of different nutritional treatments on the NO_3 content.

Shahlaei et al. (2007) evaluated the amounts of NO_3 in the edible parts of vegetables in Khuzestan province, which was produced in two different seasons (winter-spring). On the basis of NO_3 content in the vegetables, they divided it in four different groups: 1—Vegetables containing low NO_3 levels (100-500 mg kg^{-1} FW) such as potato, mint and garlic; 2—vegetables containing medium NO_3 level (500-1000 mg kg^{-1} FW) such as cauliflower, radish leaves, sweet fennel, turnip, carrot, parsley, coriander, currant, cabbage, cress, sweet pepper, eggplant, cucumber, and basil; 3—vegetables containing high NO_3 level (1000-1500 kg^{-1} FW) such as squash, root beet, celery leaves and radish tuber; and 4—vegetables containing very high NO_3 level (1500-3000 kg^{-1} FW) such as lettuce outer leaves and celery petioles. They found that (a) there was a significant difference in NO_3 content between various vegetables and season's sampling, (b) Celery petioles (3016 mg kg^{-1} FW) and potato (184 mg kg^{-1} FW) had the highest and lowest NO_3 among all the vegetables, (c) average NO_3 content in fruit bearing (tomato and cucumber), leafy and bulbous and tuberous (potato and carrot) vegetables was 999, 862, and 666 mg kg^{-1} FW, respectively, which was mostly higher than the allowable values (Table 4.1).

In the field experiments, Tabatabaei (1998), Ranjbar and Malakouti (2001), Nouri and Malakouti (2001) and Mosavifazl and Faeznia (2008) studied the effects of N fertilizers sources and rates on the yield and NO_3 accumulation in potato tubers. They found that there was a positive and significant relationship between N rates and NO_3 content. Their results were in agreement with the findings of Malakouti (2002). Supplying high rates of N fertilizer had no effect on the yield, however, NO_3 content exceeded beyond its permitted level (60 mg Kg^{-1} FW). Balanced fertilization led to the reduction in NO_3 content.

Table 4.1: Maximum, minimum and average nitrate content (mg kg^{-1} FW) of various vegetables of Ahwaz in winter and spring of 2006 and maximum tolerated nitrate contents (Shahlaei et al. 2007)*

Vegetables	Winter				Spring			
	Max	Min	Average	Allowable limit*	Max	Min	Average	Allowable limit*
Cabbage	860	28	428	900	1014	21	504	500
Lettuce inner leaves	911	300	583	4000	1522	116	862	3000
Lettuce outer leaves	2866	483	1705	4000	3172	283	1739	3000
Spinach	1400	422	886	4500	1217	483	748	3500
Radish leaves	483	55	179	500	911	178	496	400
Radish tuber	1950	608	1205	1400	5617	608	1902	1200
Cress	1406	550	679	1000	911	56	434	800
Sweet fennel	1583	178	654	1000	544	488	313	800
Basil	1460	544	996	1000	1889	438	948	800
Mint	361	55	154	1000	972	422	347	800
Coriander	910	488	569	1000	910	116	310	800
Parsley	1277	100	359	1000	1094	116	374	800
Carrot	544	238	252	1500	1217	55	458	1200
Tomato	2500	544	1644	300	2500	544	1681	200
Cucumber	2317	361	999	200	1828	178	813	150
Sweet pepper	850	300	574	500	1522	361	622	400
Potato	360	55	214	60	360	299	155	60
Currant	1094	228	446	500	605	55	385	400
Celery leaves	5127	850	2169	4000	1704	605	887	3000
Celery petioles	5829	1345	2827	5000	4293	363	2205	4000
Onion leaves	2561	667	1263	1000	577	3000	248	800
Squash	1460	100	845	500	2805	483	1471	400
Eggplant	1399	421	864	500	1399	116	652	400
Garlic	483	55	198	250	725	55	229	200
Turnip	1197	98	501	400	–	–	–	–
Red beet	1972	1142	1429	2000	–	–	–	–
Cauliflower	1032	100	555	1500	–	–	–	–
Average	1648	364	896	–	1638	281	783	–

*On the basis of first author's experience, empty allowable limits (N) have been filled.

In order to evaluate NO_3-N accumulation in potato, 60 tuber samples were collected from potato-grown regions of Khorasan Razavi Province, northeast, Iran. The results demonstrated that NO_3-N content of the samples ranged from 15 to 74 mg kg^{-1} FW with the potato yield ranging from 25 to75 t ha^{-1}. The mean NO_3-N content was 38 mg kg^{-1} FW. About 93% of the fields had NO_3-N lower than 60 mg kg^{-1} FW. They concluded that NO_3-N accumulation is lower than the permissible level in this province (Sabeghi et al. 2008). Recently, Soil and Water Research Institute (SWRI, 2010) carried out an experiment to assess NO_3 content of collected potato and onions from all over the country. The preliminary results revealed that due to existing N fertilizer subsidy and continuation of imbalanced fertilization, excess levels of NO_3 exist (Table 4.2).

Table 4.2: Nitrate content in potato and onion tubers from different locations*

Potato	Nitrate content (mg kg^{-1} FW)	Onion	Nitrate content (mg kg^{-1} FW)
Dezful-1	129	Esfahan-1	391
Dezful-2	230	Esfahan-2	615
Dezful-3	220	Esfahan-3	1218
Hamadan	201	Esfahan-4	1278
Gorghan	300	Azerbyjan-1	880
Kazeron	134	Azerbyjan-2	660

*Unpublished data (SWRI, 2010)

Irrigation and Drinking Water

Surface and groundwaters are usually polluted with NO_3. The origin of NO_3 in water, as mentioned before, is mainly from agricultural and industrial activities, i.e., overuse of N fertilizers, waste discharge, animal and human slurries. High concentration of NO_3 in water is harmful to animals and human beings, and it leads to eutrophication (the term eutrophication refers to an excessive primary production of algae and aquatic weeds caused by an increased supply of plant nutrients (Laegreid et al. 1999). Amounts of NO_3 in the underground, surface, irrigation waters and drinking water vary and mainly depend on many factors mentioned before. The amount of NO_3 in the irrigation water varies from nil (zero) to more than 100 mg L^{-1}, whereas in the natural drinking water it does not exceed 10 mg L^{-1}. Shahneazari (1996) studied NO_3 accumulation in the groundwater on paddy soils in Gilan and Mazandaran provinces during the four consecutive seasons in 1994-95 and found that NO_3 concentration in surface water, accumulating from rivers and drains, was lower than 25 mg L^{-1}, but NO_3 concentration of groundwater samples of semi-deep wells and deep-wells were more than 25 mg L^{-1} in lowland areas of the northern parts of the country. On an average, 18% house-wells had lower than 10 mg L^{-1}, 63% were within 10–25 mg L^{-1}, 9% were within 25–45 mg L^{-1}, and 5% had more than 45 mg L^{-1} NO_3, which was higher than the critical level (25 mg L^{-1}). The NO_3 concentrations in all the samples in the fall were more than the spring and summer. Most of the house-wells that had NO_3 concentrations higher than the critical level were located in the lowland paddy fields. Depths of groundwater in most cases were 0.5-5.0 meters. Distances of house-wells and septic tanks were less than 20 meters. In a few locations, the distance between the house-wells and animal stables was less than 10 meters. Most of these polluted areas were located in lowlands where overuse of urea is a common practice with shallow underground water (Malakouti 2011). Shahneazari concluded that while the concentration of NO_3 in the origin of the surface water was very low, it had increased even up to 100 mg L^{-1} after passing through villages, cities and agricultural farms due to many factors such as agricultural activities, overuse of N fertilizers, and waste and sewage materials pouring into the rivers. Such pieces of evidence were shown in shallow underground water near the seaside. Shahnazari (1996) and Malakouti (2002) believed that the overuse of subsidized N fertilizers was the main source of NO_3 accumulation in the groundwater of northern region of Iran. Malakouti (2010) in his studies on the quality of bottled mineral waters, which were usually collected from the springs, found that, on an average, NO_3 content was less than 10 mg L^{-1}. The legal limit for NO_3 in drinking water is 25–50 mg L^{-1} but in the bottled water, it should not exceed 10 mg L^{-1} (Environmental Protection Agency, 2010). However, in dry lands of Iran, i.e., Hamadan province, accumulation of NO_3 is not common in underground water due to their depth (>50 meters), and origin of NO_3 in this type of underground water cannot be attributed to agricultural activities. But, it is common in shallow underground water in agricultural lands, i.e., Gilan province (Malakouti 2002).

Cultivars

Genetic factors influencing the tissue NO_3 levels include variations in the rates of NO_3 uptake and contents or activity of NO_3 reductase. Nitrate accumulation in plants depends on the species, kind, different parts of plant and age. Vegetable crops have more susceptibility to accumulate NO_3. Generally, different species of a plant vary in absorption and decomposition of N. Usually, NO_3 accumulates in rather old parts of a plant, because in these parts the NO_3 reductive enzyme has less activity. For example, in cabbage, NO_3 accumulates in the old and mature leaves and is less in young parts.

Environmental Factors

Many environmental factors can affect NO_3 content in plants through influence on the activity of NO_3 reductive enzyme and NO_3 absorption as well. Often little light, high temperature and water stress cause the activity of NO_3 reductive enzyme to decrease and therefore NO_3 accumulation to increase. When a plant is placed in a condition with a little light or short day, the content of NO_3 increases. Malakouti (2002) found that NO_3 content in the vegetables was maximum in the early morning and least in the late afternoon. Therefore, the time of harvesting has a great effect on the amount of NO_3 that must be considered. Another environmental factor is temperature. Its effect on NO_3 accumulation is not clear, but it has been proved that an increase in temperature, especially when it corresponds with little light, leads to NO_3 accumulation. This accumulation of NO_3, which may be caused as a result of water stress, reduces the activity of NO_3 reductive enzyme and photorespiration.

Imbalanced Fertilization

Imbalanced fertilizer use is a serious threat to public health. One of the most important strategies for the promotion of national health is observing the ultimate motto of "prevention before cure and treatment of diseases". Imbalanced fertilization accelerates NO_3 accumulation in the edible parts of agricultural products. The continuity of imbalanced fertilization leads to malnutrition in human beings. The mentionable signs of malnutrition in a society include the loss of hair, irritability, chronic colds, depression, anemia, low immunity and gastrointestinal cancer. According to the World Health Organization (WHO) reports, the Iranian health ranking was 123 among 179 countries. One of the main reasons for this low health ranking in our society is imbalanced fertilization (Malakouti 2011). Continuous imbalanced fertilizer use is a serious threat to the soil health and on the performance of agricultural crops quality, particularly vegetables. Imbalanced consumption of fertilizers, particularly overuse of N and P-fertilizers has contributed to the accumulation of NO_3^- and Cd^{2+} in the soil and subsequently in the agricultural crops (Malakouti et al. 2008). The quality of vegetables and fruit, especially vitamin C content, are also reduced by continuous imbalanced fertilization practices (Welch, 2003). Currently, more than 5000 more than 7000 billion Rials subsidy is mainly dedicated to N and P-fertilizers in which their efficiencies are low due to various reasons, and secondly because these fertilizers cause NO_3^- and Cd accumulation in a variety of crops, which have a very negative role in society's health. Such a heavy subsidy has so far been allocated only to the two fertilizers mentioned, which cause farmers to show no willingness to use other fertilizers, particularly bio-organic and micronutrient fertilizers. Studies have revealed that the consumption of latter-mentioned fertilizers, in addition to being environment friendly, have very positive role in the promotion of human health. In this regard, it is suggested that N and P-fertilizer subsidies be eliminated. The types of N fertilizers should be changed for better crop production (Malakouti et al. 2008; Malakouti 2010; Malakcuti et al. 2010; Malakouti 2011).

FAO literature indicates that the soil fertility through adding fertilizers was responsible for over 55% of the growth in the agricultural outputs during the last three decades (Hamdallah 2000). At present, about 86% of the applied fertilizers in our country constitute to N and P-fertilizers. In fact, this is an alarming issue in respect to production of healthy agricultural products, and it remains a threat to food security (Malakouti et al. 2008). At present, the total amount of fertilizer use in the country is almost 4 million tons annually and from this amount 70% consists of N fertilizers, mainly urea (> 90%). The fertilizer use ratio of nitrogen (N), phosphorus (P_2O_5), potassium (K_2O), and micronutrients improved from 100-76-00+0% in 1991 to 100-44-15+1% in 1999, but it declined to 100-38-05 in 2010 due to allocating huge amounts of subsidy (700 million dollars per year) to N and P-fertilizers.

Based on the research work carried out all over the country, the annual N fertilizer needs in Iran in 2006 will be about 2.5, and in 2017 3.6 million tons. Of the total fertilizer usage estimated to be 3.4 million tons annually, more than 43% is used in wheat fields, 70% of which consists of N fertilizers, mainly urea (Welch 2003; Sanchez and Swaminathan 2005; Malakouti et al. 2008; Malakouti 2011).

NITRATE RISK ASSESSMENTS

To reduce the risk of diseases resulting from excess NO_3, determination of the permitted levels of various vegetables is necessary. In Table 4.3, on the basis of two decades of experiments, the permitted allowable levels of NO_3 content in different vegetables are shown.

Table 4.3: The permitted limit of NO_3 content in different vegetables (Malakouti 2002; Malakouti et al. 2005; Malakouti et al. 2008)

Vegetables	Nitrate content (mg kg^{-1} FW)		Vegetables	Nitrate contents (mg kg^{-1} FW)	
	Spring	Winter		Spring	Winter
Spinach	2000	2500	Kurrat	200	300
Lettuce	1500	2000	Squash	200	300
Celery petioles	1200	1500	Cucumber	150	200
Celery leaves	1000	1200	Eggplant	150	200
Radish	1000	1200	Beet root	200	200
Carrot	800	1000	Leek	200	200
Basil	600	800	Watermelon	200	–
Cress	500	700	Green been	100	100
Coriander	500	600	Pea	80	100
Parsley	400	500	Onion	60	60
Cabbage	300	400	Potato	50	50
Melon	200	–	Tomato	40	40

To assess the risk of NO_3 on human health, hazard ratio (HR) is going to be determined by the calculation of the estimated daily dose (EDD) for a human being with 75 kg body weight. In this calculation, NO_3 from vegetables, drinking water and processed meat will be estimated. For example, if a person with 66 kg body weight eats 200g vegetables with average of 900 mg kg^{-1} FW/day NO_3, 150 g potato, tomato and onion with average of 160 mg kg^{-1} FW/day NO_3, 120 g processed meat with average of 250 mg kg^{-1} FW/day NO_3 and 2L of water with average of 50 mg L^{-1} NO_3, then EDD will be calculated as follows:

EDD = (S Concentration × Quantity administered × exposure rate / Body weight

HR = (Estimated Daily Does)/Acceptable Daily Intake (3.7 mg kg^{-1} b.w./day)

$$EDD = [200 \times 900)/1000 + (150 \times 160)/1000 + (120 \times 250)/1000]/66 = 334/66 = 5.06$$

$$HR = 5.06/3.7 = 1.37$$

Since HR is more than 1.00, the consumer will face illness in the future. It is assumed that all NO_3 is absorbed by the consumer's digestive system, but, in practical, the exposure rate is less than 100%, especially in the elders.

HAZARDOUS EFFECTS OF NITRATE ON HUMAN HEALTH

Since vegetables constitute Iranian population's main side-dish their quality especially NO_3 content should be taken into consideration seriously. This is the reason why few researchers believe that most uptake NO_3 in the body is a direct result of vegetable intake. Excessive amount of NO_3 in foods and drinking water causes many problems in the human body. Young babies with low stomach acidity may suffer from infantile methemoglobinemia due to excessive NO_3 in their daily food diet, where NO_3 substitutes the oxygen in hemoglobin and death and various cancers may occur. The lethal dose of NO_3 for an adult ranges from 30 to 250 mg kg^{-1} and the average acceptable daily dose ranges from 0.4 to 13.0 mg kg^{-1} (Shahlaei et al. 2007). Permitted limit content of NO_3 in drinking water must be less than 50 mg L^{-1}. Large quantities of NO_3 in farm animals causes abortion and decrease in milk production, and in humans, especially newborn babies, it affects the growth intensively. Although, due to the shortage of water in arid and semi-arid soils, the matter of gravitational water is not sensible, this is an important issue in the northern parts of the country due to very shallow underground water. Furthermore, because of the excess application of N fertilizers, NO_3 accumulation in plants occurs. Permitted limit content of NO_3 in the edible parts of vegetables must be less than 50 mg L^{-1}. If NO_3 content in groundwater, processed foods and vegetables are exceed more than permitted limit, some disorders will happen in farm animals and humans.

The NO_3 increase in the soil caused by either the overuse of N fertilization or the nitrification process in the soil stimulates NO_3 uptake by the vegetable crops. This results in the accumulation of large amounts of NO_3 in the edible parts of vegetables. The higher content of NO_3 in the edible parts of vegetables is fatal in humans as it produces nitrosamine compounds, which have been known to be powerful carcinogens. According to the latest findings, northern region in Iran, especially Golestan province, has one of the highest rates of oesophageal cancer (OC) in the world (Akbari et al. 2006 and Islami et al. 2009). In Gonabad-Golestan, the annual incidence rates of OC are about 1.1/1000 per year for men and 1.7/ 1000 per year for women (Akbari et al. 2006). Golami (1993), in his studies on the stomach cancer patients in Sari-Mazandaran, mentioned that the incidence of gastroesophageal cancer is increasing year by year. But he did not mention the role of NO_3 as etiology of gastroesophageal cancer. He concluded that existing impurities in agricultural soils, irrigation water and drinking water could be the etiology of gastroesophageal cancer. At the end of his scientific report, he suggested that for those beyond 40-year-old, the test of gastroesophageal endoscopy is necessary in 6-month intervals. Malakouti and his colleagues (2002), in their broad investigations (Malakouti, 2005 and 2008), studied the impurities of surface, underground and drinking water of Gilan and Mazandaran provinces for two consecutive years and found that the concentration of NO_3 in both irrigation and drinking water, especially in the lowlands of paddies, is much more than its critical level. He suggested that one of the main risk factors for the incidence of gastroesophageal cancer maybe due to the practice of imbalanced fertilization, contributing to the production of unhealthy agricultural produce with high NO_3 concentrations in the edible parts of vegetables and fruits, as well as in drinking water. He summarized that (a) balanced fertilization should be practiced in all agricultural lands; (b) number of split application of urea should be increased; (c) application of urea should be eliminated and sulfur-coated-urea (SCU)+Zn be substituted with pre-plant urea in this region.

While the superiority of SCU fertilizer over pre-plant urea has been proven, especially in the light-textured soils, it is highly recommended that this substitution should be done in all types of crops, especially in the paddies (Malakouti et al. 2008).

Islami et al. (2009) have investigated the epidemiological patterns and causes of OC in this area. They have mentioned that several risk factors for OC such as cigarette and hookah smoking, opium consumption, drinking hot tea, poor oral health, low intake of fresh fruit and vegetables, and low socioeconomic status have been associated with higher risk of OC in Golestan. They have also proposed other factors such as polycyclic aromatic hydrocarbons, N-nitroso compounds, drinking water contaminated with mycotoxins, and genetic factors for further investigation as risk factors. In reference to the question of relationship between OC and NO_3 in the inhabitants of northern region of Iran, one of the authors of this paper said, "We are working on the hypothesis of the role of NO_3 as the etiology of gastroesophageal cancer and gastroesophageal reflux", but he also suggested that for those beyond 50-year-old, test of colonoscopy would be necessary in 5-year intervals (Malekzadeh,1 2011)*. Recently, in early March 2011, Iranian Aftab Newspaper announced that at present 70,000 people are suffering from cancer, mainly gastroesophageal cancer, and the main reason could be consuming unsafe foods and drinks, which contain the contaminants beyond the levels of allowable limits (critical level), at first step NO_3 and then pesticides residues and Cd in the edible parts of agricultural products (Malakouti 2011).

Since there is a positive relation between the incidence of cancer in people and unhealthy soils (Sanchez and Swaminathan 2005; Malakouti et al. 2008), another hypothesis for the existence of 70,000 cancer-affected people in our society, besides NO3 contamination, can be existing higher concentrations of pesticides and Cd in the agricultural products. At present, annually 3500 tons of pesticides are used in 230,000 hectares of paddies, just in Mazandaran province. On the other hand, herbicides are the most important soil and water pollutants in paddy fields. Herbicides translocation to the surface and groundwaters and to the food chain is a major concern for the environment and human health. It is, therefore, necessary to investigate the fate of these pollutants and their reactions in natural ecosystems. To meet this purpose, Mahmoudi (2010) investigated the residual effect of pesticides on rice, spinach and lettuce in Mazandaran, and found that thiobencarb and oxadiargyl concentrations were reduced in soil solution rapidly and reached a

* Personal communication with Professor R. Malekzadeh, to whom we herewith extend our acknowledgement

steady state after approximately 30 days and 20 days, respectively. Calculations revealed that thiobencarb and oxadiargyl are persistent herbicides in the soil due to their strong adsorption and high hydrophobic coefficients. Soil characteristics, in particular organic matter content, were the main factors controlling the effect of thiobencarb and oxadiargyl residues. Experimental data also demonstrated that thiobencarb and oxadiargyl were taken up by rice, spinach and lettuce and stored in the shoots. Increase in thiobencarb and oxadiargyl applied doses led to an increase in their concentration in spinach and lettuce leaves, which was hazardous to human health. While the concentration of thiobencarb in the control plots (soil, rice components, spinach and lettuce) was almost zero, it increased in soil after 3-month application, 0.34 mg kg^{-1}, in white rice grain 0.13, in bran 0.31 and in rice straw 0.30, in spinach (after 6 months) 150 and in lettuce (after 9 months) exceeded 0.30 mg kg^{-1} (Mahmoudi 2010). The concentrations of thiobencarb and oxadiargyl in rice components, spinach and lettuce leaves are shown in Figs. 4.8, 4.9, 4.10 and 4.11.

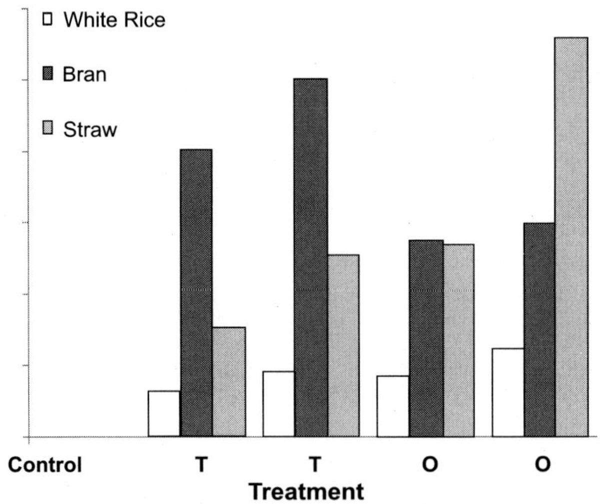

Fig. 4.8. Residual effects of herbicides on rice components (Mahmoudi 2010).

We strongly believe that there is a positive link between gastroesophageal cancer and unsafe foods, fast foods and drinks, which contain contaminants beyond levels of allowable limits (Malakouti 2011).

STRATEGIES FOR REDUCING NITRATE IN VEGETABLE CROPS

Since N fertilizers are the main sources of NO_3 accumulations in the agricultural products, and unfortunately more than 90% of the N fertilizers being used in our country is urea, so substituting SCU with pre-plant urea in all irrigated crops, especially in lowland agricultural regions and increasing number of split applications of urea in top dressing application (recommending fertigation practice), should be taken into consideration (Malakouti et al. 2008). Among other factors which are

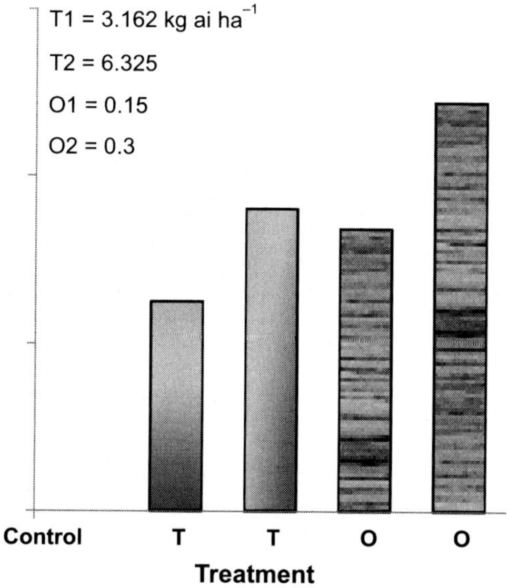

Fig. 4.9. Residual effects of herbicides on white rice grain (Mahmoudi 2010).

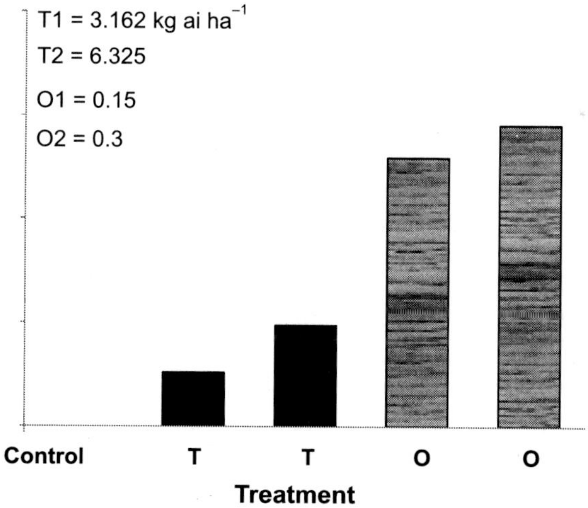

Fig. 4.10. Residual effect of herbicides on spinach leaves (Mahmoudi 2010).

very important in reducing NO_3 content in the consumed foods is practicing balanced fertilization (One of the main principles of sustainable agriculture is to return to the soil those nutrients removed from it by the crops). By practicing balanced fertilization, applying zinc sulfate and bio-organic fertilizers, the country's nutrition will improve and the quality and quantity of agricultural production will be upgraded significantly, and hence the problem of malnutrition will be solved. On the other hand, by implementing optimal fertilizer management practices, crop

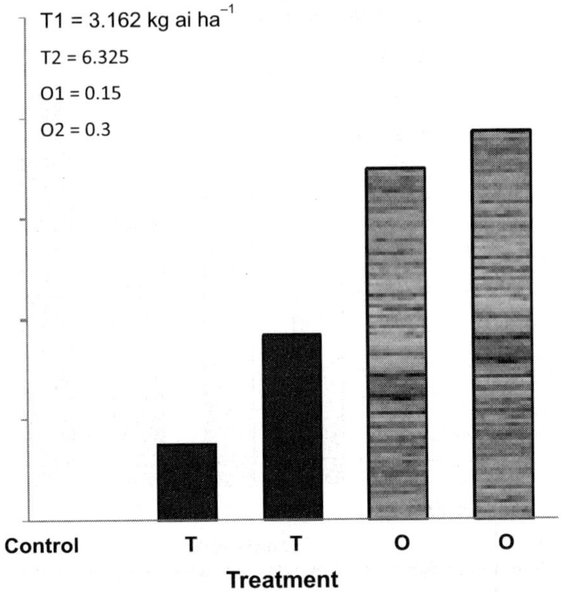

Fig. 4.11. Residual effect of herbicides on lettuce leaves (Mahmoudi 2010).

enrichment will be accomplished in farms and orchards, which in addition to achieving sustainable production of healthy crops, will also upgrade the health rating of Iranian society. Therefore, at this time, scientific management of production and application of all kinds of fertilizers is inevitable. In this case, by improving the quality of all types of fertilizers, and following the principle of balanced fertilization, besides achieving the desired yield increase, the quality of agricultural products will be enhanced and the level of society's health will be promoted automatically. Furthermore, there is a close and positive relationship between balanced fertilization and production of healthy agricultural products.

After 20-year studies and investigations it has been proven that (a) there is a link between soil quality and health of human beings, and human malnutrition is mainly due to low soil quality management; (b) health comes from the farm, not from the pharmacy; enrichment of the agricultural products should be taken seriously into considerations by the Government Policy Makers, and it is obvious that it should be emphasized in making practical of this cheapest and applicable method of crop enrichment from the farm which its superiority over food fortification as well as supplements are clear (Malakouti 2011). Proposed ideal fertilizer use ratio in calcareous soils should be 100-35-30+35(S-fertilizers) + 4% (micronutrients). In this regard, the amount of N and P-fertilizers should be reduced, and instead, the consumption of complete fertilizers, SCU, zinc sulfate, S and biofertilizers should increase, especially in pre-plant applications to obtain higher efficiencies. Proper distribution of subsidized fertilizers, improved management and balanced nutrition should also be taken to practice. Fertilizer recommendations should be done based on the results of soil, plant, and water and fruit analysis. Quality control of agricultural products should be done according to international standards (Malakouti et al. 2008; Malakouti 2010).

Growers can play a key role in reducing NO_3 content in the vegetable crops. One of the strategies to decrease NO_3, is the use of cultivars, which accumulate less NO_3 in comparison with other cultivars. In general, light intensity has an important role in NO_3 assimilation, so any factor that reduces light interception should be avoided. For example, cultivation of a number of plants together that cause plants to shade vertically intensifies NO_3 accumulation. In glasshouse production crops, the use of artificial light during night before harvesting would be useful in lowering NO_3 accumulation in vegetables. The nutrient supply in proportion to the vegetables daily uptake, split application of N fertilizers, balanced fertilization, SCU application are also among the strategies to reduce NO_3 in the edible parts of vegetables. Moreover, water stress must be prevented. Vegetable consumer should notice that the petiole of leafy vegetables, especially spinach, lettuce and celery has a large amount of NO_3, therefore, they should be produced in a scientific way. In vegetables such as carrot and beet root, maximum amount of NO_3 is accumulated in the end parts, and in some leafy vegetables such as lettuce and cabbage, the external aged leaves have much more NO_3, so they are not suitable to be eaten. Application of a large quantity of N fertilizers, especially urea, in fall cultivation should be avoided. Sowing time with an appropriate density and harvesting the crops in afternoon promotes NO_3 reduction. Nitrate accumulation occurs in dark period with a little amount of light. Therefore, if crops are harvested in afternoon, NO_3 content in comparison with morning harvest will be much less. Also harvesting in sunny days results in less NO_3 content than cloudy days. Peeling, cooking and washing before and after using, application of balanced fertilization, especially K, Zn and B, are other ways to reduce NO_3 content in the edible parts of vegetables (Malakouti et al. 2004, 2005; Malakouti et al. 2008).

Recently, some researchers introduced a new technique in which by using fine structure Fe^0 nanoparticles they decontaminated NO_3 in wastewater and groundwater. In this method, Fe^0 releases electron and NO_3 is reduced to N_2 or NH_3 while the oxidizing Fe converts to Fe_2O_3 or Fe_3O_4 electrochemically. The efficiency of this reaction is very high in low pH (Lin et al. 2008). Mikhak (2010) studied nitrate decontamination using quarts supported Fe^0 nanoparticles and found that nitrate reduction increased with an addition to initial concentration of Fe^0 nanoparticles and ionic strength; however, an adverse relation was found for initial pH. Fe^0 nanoparticles made by electric arc were more effective in NO_3 removal in spite of their higher size. This was attributed to the possible aggregation of Fe^0 nanoparticles made by chemical method during the production process. Aggregation reduces the reactive surface area. This was somewhat eliminated by coating Fe^0 on sand particles. Coating enhanced NO_3 reduction by 25%. This was attributed to a better reaction of Fe^0 with NO_3 by increasing the effective surface area. Based on the results, it may be concluded that Fe^0 nanoparticles can be effectively used for NO_3 reduction from water resources, and coating process increases the reactive surface area and therefore the efficiency NO_3 removal. With respect to the importance of this research in human health, it is recommended to evaluate this technique on a large-scale dimensions from the viewpoints of its feasibility and economics (Kassaee et al. 2011).

CONCLUSION

In the third millennium, the promotion of human health should be the priority of the crop producers. So, balanced fertilization should be seriously practiced on the basis of soil and plant analysis data. The accumulation of NO_3 in the edible parts of vegetables depends on various factors, especially the amounts, timing and sources of N fertilizers and cultivars. Therefore, decreasing these types of contaminants is necessary for the promotion of human health. In the farmers' conventional method of fertilizer use (overuse of N and P-fertilizers due to existing heavy subsidy), shelf-life of vegetables and vitamin C content were decreased due to higher percentage of water. Imbalanced fertilization results in higher NO_3 content in the edible parts of vegetables. Increasing K and Zn-fertilizers is associated with their highest yield and better quality. Nitrate is a naturally occurring compound that is a part of N cycle and plays an important role in the nutrition and metabolic function of plants. However, it is possible to minimize the NO_3 content by applying good agricultural practices in producing the vegetables. The following approaches have been summarized to minimize the risk of NO_3 contamination:

- Continuous imbalanced fertilizer use is a serious threat to the soil health and the quality and quantity of agricultural crops, particularly vegetables. Imbalanced consumption of fertilizers, particularly overuse of N fertilizers, has contributed to the accumulation of NO_3^- in the soil and subsequently the agricultural crops. The quality of vegetables and fruits, especially vitamin C content, is also reduced. In this regard, it is suggested that N fertilizer subsidy should be eliminated.
- There is a large variation in the content of NO_3 in different vegetables from 20 mg kg^{-1} FW to 4000 mg kg^{-1} FW; leafy vegetables have been shown consistently to have the highest levels, especially in their outer leaves when harvested in the early morning under high N fertilizer application situation.
- Nitrogen from both organic as well as inorganic fertilizers is the main source of NO_3 in the soil, therefore, excessive and untimely N fertilization increases NO_3 accumulation in the plants.
- Nitrate predominantly enters the human body exogenously from vegetables, drinking water, processed and fast foods. Daily use of different types of vegetables, foods and drinks plays an important role in the daily intake of NO_3.
- Some vegetables, particularly leafy vegetables, have been shown to have relatively high levels of NO_3, but implications for food safety are unclear. In order to provide a scientific basis for a long-term strategy to manage any risks from dietary NO_3, an updated risk assessment is considered necessary by the government.

- Light intensity, storage, processing and/or cooking of vegetables affect the NO_3 content.
- Delaying harvest until afternoon is advisable, as it results in a significant reduction in tissue NO_3 levels.
- There is a need for more work into the factors that influence NO_3 content in the vegetables crops and alterations during productions, storage and processing.
- Fertilizer recommendations should be done based on the results of soil, plant, and water and fruit analysis. Quality control of agricultural products should be done according to international standards. By practicing balanced fertilization, the country's nutritional ranking will improve and the yield and quality of crops, especially tastes and shelf-life of vegetables will be upgraded significantly, and hence the problem of malnutrition will be solved. Finally, the authorities of the Ministry of Jehad-e-Agriculture are supposed to consider this important principle for (a) supplying suitable kinds and amounts of N fertilizers during application seasons in producing agricultural products; (b) banning the act of burning the plants residues; (c) considering the rules of balanced fertilization; (d) enrichment of agricultural products on the farm; (e) scientific management of production and application of all kinds of fertilizers; and (f) removing the subsidy on N and P-fertilizers in order to increase the yield and quality of agricultural products and enhance health of the society. By practicing optimal management in fertilizer use, enrichment of agricultural products in farms, and using healthy agricultural products, the ranking of the Iranian society's health will be upgraded.

ACKNOWLEDGEMENTS

We would like to thank all the researchers and staff members of Soil and Water Research Institute from all over the country, as well as all those graduate students of Soil Science Department of Tarbiat Modares University who have made valuable contributions to this paper. We also wish to thank Professor M. Bybordi and Professor Reza Malekzadeh, Mr. H. Amirmokri, Ms. T. Hedayati for their valuable assistance.

REFERENCES

Akbari, M.R., Malekzadeh, R., Nasrollahzadeh, D., Amanian, D., Sun, P., Islami, F., Sotoudeh, M., Sh Semnani, Boffeta, P., Dawsey, S.M., Ghadirian, P. and Narod, S.A. (2006). Familial risks of esophageal cancer among the Turkmen population of the Caspian littoral of Iran. *Int. J. Cancer* **119:** 1047-1051.

Anac, D. and Martin-Preve, P. (1999). Improved crop quality by nutrient management (Eds.). Kluwer Academic Publishers. London, England.

Andersen, L. and Nielsen, N.K. (1992). A new cultivation method for the production of vegetables with low content of nitrate. *Scientia Hort.* **49:** 167-171.

Anjana, Umar, S. and Iqbal, M. (2006). Nitrate accumulation in plants, factors affecting the process, and human health implications. *Agron. Sustain Dev.* **26:** 1-13.

Anjana, Umar, S., Iqbal, M. and Abrol, Y.P. (2006). Are nitrate concentrations in leafy vegetables within safe limits? *Curr. Sci.* **92:** 355-362.

Bakhteyari, Sh, (2005). Analysis of yield and nitrate content in hydoponic vs conventional cucumber (M.Sc. thesis). Faculty of Plant Biology. Payam-e-Nour University. Tehran, Iran (in Farsi).

Bureau of Statistics and Information Technology Office, (2007). Agriculture Statistics. Vol. 1: Agronomic and horticultural crops data for 2005-06. The Agency for Planning and Economics. Ministry of Jahad-e-Agriculture. E-mail: Info@agri-Jahad-org; http//www.maj. ir; Tehran, Iran (in Farsi).

Bybordi, A., Tabatabaei, S.J., Malakouti, M.J., Basirat, M. and Rajabzadeh, F. (2004). Study on the nitrate variation among different vegetables. pp. 145-156. *In:* Malakouti, M.J. Bybordi, A. and Tabatabaei, S.J. (Eds.). Balanced fertilization of vegetable crops: An approach to improve yield and quality of vegetables, reduce contaminants and enhancing human health (A compilation of papers). Agronomy Dept. Ministry of Jihad-e-Agriculture (in Farsi).

Environmental Protection Agency, (2010). Nitrate in drinking water. www.epa.ie

Golami, M. (2004). Study on the status of gastric and esophageal cancer patients in Sari-Mazandaran province. Bull. No. 3. Mazandaran Medical Sciences University. Sari, Iran (in Farsi).

Hamdallah, H. (2000). Soil fertility management: The need for new concepts in the region. Soil Fertility Management through Farmers Field School in the Near East Countries. FAO, Amman, Jordan.

Hashemimajd, K. and Fathi, B. (2006). Estimation of the amounts of nitrate daily intake by the consumers in Ardabil province (Abstarct, p. 37). The 2nd Conference and Grand Exhibition of Food Industries. Isfahan Technology University. Isfahan, Iran (in Farsi).

Heffer, P. (2008). Assessment of fertilizer use by crop at the global level. International Fertilizer Industry Association. rue Marbeuf, Paris, France. www.fertilizer.org. p. 5.

Islami, F., Kamangat, F., Nasrollahzadeh, D., Moller, H., Boffetta, P. and Malekzadeh, R. (2009). Oesophageal cancer in Golestan Province, a high-incidence area in northern Iran–A Review. *European J. Cancer* **45:** 3156-3165.

Kessaee, M.Z., Motamedi, E., Mikhak, A. and Rahnemaie, R. (2011). Nitrate removal from water using iron nanoparticles produced by arc discharge vs. reduction. *Chem. Eng. J.* **166:** 490-495.

Kumar, A., Martin, A., Parry, J., Rowan, A., Mitchell, C., Ahmed, A. and Abrol, Y. (2002). Photosynthesis and nitrogen use efficiency. *In:* Foyer, C.H. and Noctor, G. (Eds.). Kluwer Academic Publishers. **12:** 23-34.

Laegreid, M., Bockman, O.C. and Kaarstad, O. (1999). Agriculture, fertilizers and the environment. NorskHydro ASA. CABI Publishing, Norway. p. 294.

Lin, K.S., Chang, N.B. and Chuang, T.D. (2008). Fine structure characterization of zero-valent iron nanoparticles for decontamination of nitrites and nitrates in wastewater and groundwater. Science Technology of Advanced Materials, p. 9.

Mahmoudi, F. (2005). Investigation of the effects of amount and sources of N fertilizers on the nitrate accumulation and yield of lettuce (M.Sc. thesis). Soil Science Dept., Islamic Azad University. Tehran, Iran (in Farsi).

Mahmoudi, M. (2010). Adsorption and kinetics of oxadiargyl and thiobencarb degradation in calcareous paddy soils. Ph.D. dissertation. Soil Science Dept., Tarbiat Modares University. Tehran, Iran., p. 126 (in Farsi).

Malakouti, M.J. (1996). Sustainable agriculture and yield increase through balanced fertilization. Ministry of Agriculture. Karaj, Iran., p. 350 (in Farsi).

Malakouti, M.J. (2000). Nitrate levels in potatoes, onions, and edible parts of vegetables affect human health (Technical note no.1). *Iran J. Soil Water Sci.* **12:** 1-5.

Malakouti, M.J. (2002). Evaluation of the effects of N fertilizers on nitrate accumulation in the edible parts of vegetables (Final report). Agricultural Commission, National Council for Scientific Research. Tehran-Iran., p. 43 (in Farsi).

Malakouti, M.J. (2002). Study on the origin and methods of decreasing nitrates and cadmium contaminants in the paddy soils of northern Iran (Final report). Agricultural Commission, National Council for Scientific Research. Tehran-Iran., p. 43 (in Farsi).

Malakouti, M.J. (2005). The trends in N fertilizer use and the necessity for increasing nitrogen use efficiency (NUE) in the calcareous soils of Iran. 1st International Iranian Urea/Ammonia Conference, Ministry of Oil. Tehran, Iran.

Malakouti, M.J. (2008). Balanced fertilization is the most appropriate way for sustainable agriculture and solving society's cell hunger. Agro Ecology College, Shahid Beheshti University. Tehran, Iran (in Farsi).

Malakouti, M.J. (2011). Relationship between balanced fertilization and healthy agricultural products: a review. Tabriz Islamic Azad University. *Agric. Sci. J.* (in press).

Malakouti, M.J. (2011). Practical methods for improving the quality of consumed breads in Iran: A review. *Iran J. Agric. Sci.* (under consideration)

Malakouti, M.J. and Mashayekhi, H.H. (1995). Necessity for the use of zinc sulphate in improving yield, quality and enriching agricultural products in the country. Technical Bull. No. 25, Ministry of Agriculture. Tehran, Iran., p. 16 (in Farsi).

Malakouti, M.J. and Tabatabaei, S.J. (1996). Potato yield increase and its NO_3 control through appropriate use of fertilizers. Technical bulletin No. 150. Ministry of Agriculture. Tehran, Iran., p. 18 (in Farsi).

Malakouti, M.J. and Bybordi, A. (2006). Interaction between potassium and zinc on the yield and quality of tuber vegetables. Proceeding of the Balanced Fertilization for a Sustaining Crop Productivity. Ludhiana, India.

Malakouti, M.J., Bybordi, A. and Tabatabaei, S.J. (2004). Balanced fertilization of vegetable crops: An approach to enhance the yield and quality of vegetables, reduce contaminants and improve human health. Agronomy Dept., Ministry of Jihad-e-Agiculture. Tehran, Iran. p. 338 (in Farsi).

Malakouti, M.J., Shahabi, A.A. and Bazargan, K. (2005). Potassium in Iranian agriculture. Sana Publication Co., Ministry of Jihad-e-Agriculture. Tehran, Iran. p. 294 (in Farsi).

Malakouti, M.J., Keshavarz, P. and Karimian, N. (2008). A comprehensive approach towards identification of nutrient deficiencies and optimal fertilization for sustainable agriculture. Tarbiat Modares University Press. Tehran, Iran. p. 755 (in Farsi).

Malakouti, M.J., Nouri, O., Samavat, S. and Basirat, M. (2005). Reasons for accumulation of nitrates in the edible parts of fruit-vegetables. Pub. No. 414. Ministry of Jihad-e-Agriculture. Tehran, Iran., p. 18 (in Farsi).

Malakouti, M.J., Bybordi, A., Ranjbar, R. and Nouri, A.A. (2007). The role of zinc (Zn) and potassium (K) in the reduction of nitrate (NO_3) and cadmium (Cd) contaminants in potato and onion (Short paper). Zinc Crops 2007: Improving crop production and human health Conference. Istanbul, Turkey.

Malakouti, M.J., Malakouti, A., Bybordi, I. and Khamesi, E. (2010). Zinc (Zn) is the neglected element in the life cycle of plant, animal and human health (10th edition with complete revision). Technical bulletin No. 007. Soil Science Dept., Tarbiat Modares University. Sana Publication Co., Tehran, Iran. p. 14 (in Farsi).

Malakouti, M.J., Bybordi, A., Lotfollahi, M., Shahabi, A.A. and Siavoshi, K. et al. (2008). Comparison of complete and sulfur coated urea fertilizers with pre-plant urea in increasing grain yield and nitrogen use efficiency in wheat. *J. Agric. Sci. Tech.* **10:** 173-183.

Malekzadeh, R. (2011). The role of nitrate as etiology of gastric, esophageal cancer and gastro esophageal reflux. Tehran Medical Sciences University, Tehran-Iran (personal communications).

Marschner, H. (1995). Mineral nutrition of higher plants. Second ed., Academic Press. Harcourt Brace Company, Pub. Co. New York. p. 890.

Mikhak, A. (2010). Nitrate decontamination using quarts supported Fe^0 nanoparticles (M.Sc. thesis). Soil Science Dept., Tarbiat Modares University. Tehran-Iran., p. 53 (in Farsi).

Mousavifazl, S.H. and Faezniya, F. (2008). Effect of water and N fertilizer on the yield and nitrate concentration of potato tubers. *Iran J. Soil Water Sci.* **22:** 243-250 (in Farsi).

Nouri, A.A. and Malakouti, M.J. (2001). Effect of potassium and zinc sulfate fertilizers on increasing potato yield and decreasing its cadmium and nitrate contents in Zanjan's heavy-textured soils. *Iran J. Soil Water Sci.* **12:** 104-109 (in Farsi).

Office of Health and Environmental Assessment, (1990). Integrated risk information system: Last review. Toxicological Information on nitrate report. USEPA. WA. DC. USA.

Pishnamaz, F. (2005). Improving nitrogen use efficiency and minimizing nitrate accumulation with high yielding potato in Zanjan province (M.Sc. thesis). Soil Science Dept., Islamic Azad University. Tehran, Iran (in Farsi).

Rajabzadeh, F. (2004). Variation in nitrate levels of vegetables in Tehran markets (M.Sc. thesis). Soil Science Dept., Islamic Azad University. Tehran, Iran (in Farsi).

Ranjbar, R. and Malakouti, M.J. (2001). Effects of rates and sources of potassium fertilizers and zinc sulfate on the nitrate and cadmium levels in potato in east calcareous soils (Part 2). *Iran J. Soil Water Sci.* **12:** 121-127 (in Farsi).

Raun, W.R. and Johnson, G.V. (1999). Improving nitrogen use efficiency for cereal production. *Agronom. J.* **91:** 357-363.

Sabeghi, H., Abbaszadeh, M. and Zarifian, R. (2008). Investigation of NO_3-N accumulation in potato. Proceedings of Potato Future Food Symposium. Sarvelayat- Khorasan, Iran (in Farsi).

Sanchez, P.A. and Swaminathan, M.S. (2005). Hunger in Africa: The link between unhealthy people and unhealthy soils. *Lancet* **365:** 442-444.

Shahlaei, A., Ansari, A.A. and Dehkordie, F.S. (2007). Evaluation of nitrate and nitrite content of vegetables in Khuzestan, Iran. *Asian J. Plant Sci.* **6:** 1197-1203.

Shahnazari, R. (1996). Study on the nitrate accumulation in the groundwater of paddy soils in Gilan and Mazandaran provinces. Soil Sci. Dept., Tarbiat Modares University. Tehran, Iran.

Six, J. (2011). Plant nutrition for sustainable development and global health (editorial). *Plant Soil* **339:** 1-2.

Tabatabaei, S.J. (1998). Effects of balanced fertilization on nitrate reduction in potato in Ajabshir region (M.Sc. thesis). Horticultural Dept., Tarbiat Modares University. Tehran, Iran (in Farsi).

Welch, R.M. (2003). Farming for nutritious foods: Agricultural technologies for improved human health. IFA-FAO Agricultural Conference on Global Food Security and the Role of Sustainable Fertilization. Rome, Italy.

5

Studying Nitrate Accumulation in Spinach and Cabbage under Organic Conditions

Sahriye Sonmez and Sedat Citak

Akdeniz University, Faculty of Agriculture, Soil Science Department, Antalya/Turkey
e-mail: ssonmez@akdeniz.edu.tr

ABSTRACT

Organic farming has emerged as an outcome of consumer's reaction against harmful toxins and the desire for healthier life and safe environment. Not only does organic farming supply healthy foods to human beings, but also conserves the environment in which agricultural practices occur. Vegetables occupy an important role in human diet and are consumed in large quantities. However, vegetables, especially leafy vegetables, could contain some undesired compounds that are hazardous for human health like nitrate, being toxic to human in higher concentrations. Spinach and cabbage are two of the most important leafy vegetables in terms of their cultivation area and consumption rate in the world. This review focuses on the elaboration of nitrate accumulation in organically grown spinach and cabbage.

Keywords: Organic farming, Nitrate, Spinach, Cabbage

INTRODUCTION

One of the most important roles of agricultural production is to provide almost all essential minerals and organic nutrients to humans (Wang et al. 2008). Organic agriculture is a production system that sustains health of soils, ecosystems and people. It relies on ecological processes, biodiversity and cycles adapted to local conditions, rather than the use of external inputs with adverse effects. Organic agriculture combines tradition, innovation and science to benefit the shared environment, and promotes fair relationships and a good quality of life for all involved (Anonymous 2007). According to the latest survey on organic farming worldwide, almost 37.3 million hectare areas and more than 1.8 million farms were managed organically in 2009 (SOEL 2011).

Organic agriculture has made a steady increase both in acceptance and popularity in recent years. This increase is being enhanced by shifting consumers' demand in favor of health concerns. Also, Warman and Havard (1997), Ismail and Fun (2003) stated that many consumers perceive that organically-grown foods are better, healthier and more nutritious than conventional counterparts.

NITRATE ACCUMULATION IN PLANTS

Nitrate is the natural constituent in plants and is present in all vegetables, with higher levels commonly found in green vegetables such as spinach, lettuce and celery, and root vegetables such as beetroot and radish (Anonymous 2006). Nitrates are readily taken up by crops, and if they are not utilized immediately in the formation of protein, they are stored in the cells in their original form (Lampkin 2001).

Nitrate content is an important quality characteristic of vegetables (Santamaria 2006). Pesticide residues is not the only problem arising from modern agricultural techniques; nitrate levels in vegetables are also causing concern (Lampkin 2001). Nitrate is ubiquitous and is present in food and water; it is also a normal human metabolite. There is no single source of nitrate in the body. Food and water are the main external sources of nitrate in human body, vegetables account for more than 80% of nitrate intake, and water and other foods account for less than 10% each (L'hirondel 2001). Lampkin (2001) pointed out that 70% of average daily nitrate intake comes from vegetables, compared with only 20% from drinking water.

Leafy vegetables such as cabbage, spinach, lettuce, and celery provide many kinds of vitamins and mineral elements to humans. Since these crops are usually harvested at vegetative growth stages, and the edible parts — young stems and leaves — can accumulate relatively large amounts of nitrate, leafy vegetables have been found to be the major source of nitrate uptake by humans (Wang et al. 2008).

Leafy vegetables are the main part of the human diet (Anjana et al. 2007). However, these vegetables may contain excessive levels of nitrate due to application of excessive amount of nitrogenous fertilizers to achieve high yield and to appeal the appearance, especially in winter season.

Santamaria (2006) pointed out that the vegetable nitrate content is of interest to government and regulators owing to the possible implications for health. On 8 November 2005, the European Commission issued a regulation stating the maximum acceptable nitrate concentration in lettuce and spinach. For spinach the following limits are fixed: 3000 and 2500 mg kg^{-1} on fresh weight basis for crops harvested from November 1 to March 31 and from April 1 to October 31, respectively (Anonymous 2005).

The acceptable daily intakes (ADI) of nitrites and nitrates recommended by WHO are 0–0.06 mg NO_2 kg^{-1} and 0–3.7 mg NO_3 kg^{-1} body weight (Anjana and Iqbal 2007; Anonymous 2008). Therefore, assuming a 60 kg body weight, ingestion of only 100 g of fresh vegetables with a nitrate concentration of 2500 mg kg^{-1} fresh weights exceeds the Acceptable Daily Intake for nitrate by approximately 13% (Anjana and Iqbal 2007).

According to Majumdar (2003), it is of importance to manage the nitrate level in vegetables in terms of some diseases including Blue Baby syndrome. European Food Safety Authority states that nitrate *per se* is relatively non-toxic, but its metabolites and reaction products, e.g., nitrite, nitric oxide and N-nitroso compound, have raised concern because of implications for adverse health effects such as methemoglobinemia and carcinogenesis (Anonymous 2008). Blue Baby syndrome or methaemoglobinemia is caused by decreased ability of blood to carry oxygen, resulting in oxygen deficiency in different body parts (Majumdar 2003). Some signs and symptoms of methaemoglobinemia are given in Table 5.1.

Table 5.1: Some signs and symptoms of methemoglobinemia

Methaemoglobinemia Concentration (%)	Clinical Findings
10–20	Central cyanosis of limbs/trunk; usually asymptomatic
20–45	Central nervous system depression (headache, dizziness, fatigue, lethargy, syncope), dyspnea
45–55	Coma, arrhythmias, shock, convulsions
>70	High risk of mortality

Source: From Anjana and Iqbal 2007

Ascorbic acid (vitamin C) can afford protection from this conversion of nitrate to nitrite to nitrosamines (Mirvish et al. 1972), and it is of some interest that the higher levels of nitrate in non-organic crops have often been linked with lower ascorbic acid levels (Rembialkowska 2000). Additionally, Hornick and Parr (1990) showed that increasing levels of fertilizers increased the yields, but decreased the ascorbic acid content of the crops. Furthermore, Schudel et al. (1979) found that organically grown spinach contains more vitamin C and fewer nitrates.

NITRATE ACCUMULATION IN SPINACH AND CABBAGE

Spinach is the most important leafy vegetable grown and consumed all over the world. Based on the FAO statistics (Anonymous 2013), spinach production occupied 867.851 ha area and was produced 21.005.505 tons; and cabbage and other brassicas are also important leafy vegetables and was consumed over the world, having 2.373.818 ha production area and 68.840.531 tons of total production in 2011.

In addition to essential minerals, spinach also contains vitamins and antioxidants. Bhattacharjee et al. (1998) stated that spinach is a mineral-rich vegetable and is also a good source of vitamin B complex, ascorbic acid, vitamin A and carotin. Jablonska-Ceglarek and Rosa (2003) stated that cabbage has considerable nutritious value, high yield capacity from a unit area, and the possibilities of utilization throughout the year.

The factors responsible for nitrate accumulation in plants are mainly nutritional, environmental and physiological (Anjana and Iqbal 2007). Nitrate accumulation in vegetables is caused due to the imbalance between nitrate absorption and reduction by plants (Wang et al. 2008). Higher levels of nitrate are present in leaves, whereas lower levels occur in seeds or tubers. Thus, leaf crops such as lettuce and spinach generally have higher nitrate concentrations, and human exposure to nitrate is mainly exogenous through the consumption of these vegetables, and to a lesser extent water and other foods (Anonymous 2008). Spinach is classified as a very high level nitrate containing vegetable accumulating more than 2500 mg kg^{-1} Fresh Weight (FW). Cabbage contains lower level of nitrates than the spinach, and accumulates 500–1000 mg kg^{-1} FW nitrate. It is classified as a medium level nitrate containing vegetable (Santamaria 2006).

Organic manure management is a key point in controlling the nitrate accumulation in plants in organic agriculture. One should keep in mind that organic manure to be used in the production should be composted before the application. Schultz et al. (1992) showed that fresh manure application resulted in higher nitrate content in carrots than the composted manure. Selecting the manure type is also important for managing the nitrate content in organic crops. In Table 5.2, effects of different farming systems and seasons on nitrate content of the spinach is given.

Table 5.2: Nitrate content in fresh weight spinach samples (Moramoto, 2000)

Seasons	Practices	n	NO_3 mg/kg whole plant FW			
			Average	Minimum	Maximum	CV (%)
Winter	Conventional	6	2,230	1,500	2,900	25
	Organic	6	1,800	890	2,600	34
Winter total		12	2,010	890	2,900	30
Summer	Conventional	6	2,850	2,000	3,400	17
	Organic	6	1,820	600	3,000	61
Summer total		12	2,330	600	3,400	42
Conventional total		12	2,450	1,500	3,400	23
Organic total		12	1,810	600	3,00	47
Grand total		24	2,170	600	3,400	37

From Table 5.2, it can be concluded that organic spinach grown especially in winter season tends to accumulate lower nitrate than the summer season and conventional counterparts. According to Fujiwara et al. (2005), organically grown spinach in winter season contains lower level nitrates than in the summer season. Rosen and Allan (2007) stated that the most important problem in organic agriculture is the imbalance between the mineralization rate of manure and the crop demand. However, the mineralization rate of manure, depending on the conditions and manure type, in winter season is assumed to be 0–20% (Lampkin 2001). Citak and Sonmez (2010) studied different organic manures in the growing spinach and revealed that organic manure caused nitrate content to be lower than the conventional one; further, winter season plants contained less nitrate than that of summer season plants.

Stopes et al. (1989) found that fertilisation with farmyard manure compost (FYMC) did not significantly increase the nitrate accumulation when compared with a zero fertilized control. Moramoto (2000) pointed out that organic spinach grown using guano tends to accumulate higher nitrate levels than the spinach grown using compost. Blood meal can also be used as a manure, however, as reported by Termine et al. (1987), mineral fertilizers and blood meal cause similar effects—higher nitrate levels in vegetables. Abraham-Gutierrez et al. (2004) stated that blood meal should not be used as a single application. Leuzzi et al. (1996) observed that using blood meal in spinach production increased the nitrate level.

Demir et al. (1996) studied the effects of different doses of chicken manure and farmyard manure application on nitrate content of spinach, and found 61.9–1278.8 mg NO_3^- kg^{-1} in FW for chicken manure application, and 52.8–1428.0 mg NO_3 kg^{-1} FW for farmyard manure application. Besirli et al. (2004) found that organically grown spinach contains 54.24–171.95 mg kg^{-1} FW nitrate. Moreover, Citak and Sonmez (2010) found nitrate content of spinach to be in the range of 20.3–76.0 mg kg^{-1} FW in autumn season, and 14.1–163.2 mg kg^{-1} FW in winter season.

Ahrens et al. (1983) found lower nitrate content and higher vitamin C content in organically grown spinach at harvest, and lower nitrite formation during the storage. Baker (1975) pointed out that organically grown spinach gave lower crop nitrate contents due to slower mineralization rates.

Kopp (1992) and Citak and Sonmez (2011) studied some vegetables including cabbage and found that organically grown cabbages contain lower level of nitrate and higher level of vitamin C. Lairon et al. (1986) also found that organically grown vegetables contain lower level of nitrate. Also, Scuphan (1974) stated that organically grown vegetables including spinach are characterized by their lower nitrate levels.

Rembialkowska (2000) showed that organically grown cabbages contain 3–4 fold lower nitrate than the conventional ones, and Rembialkowska (2007) stated that cabbages grown under organic conditions contain 99–334 mg kg^{-1} nitrate in fresh weight. Fytianos and Zarogiannis (1999) also reported that cabbage samples contain 19.6–414 mg kg^{-1}nitrate on fresh weight basis. Zahradnik and Petrikova (2007) studied the effects of different organic manures on the nitrate content of cabbages 40-153 mg kg^{-1}.

CONCLUSION

Producing healthy foods is the main aim of organic farming practices. It should be noted that selecting the type of organic manure and growing season are important factors in maintaining the nitrate level in spinach and cabbage. The main factor responsible for releasing the nutrients from manure is mineralization rate. Therefore, the factors affecting the mineralization rate also affect the nitrate accumulation in plants depending on the conditions and plant species. In this review, farmyard manure and chicken manure appeared to be less effective in accumulation of nitrate in vegetables. Guano type and blood meal, however, appeared to more effective in nitrate accumulation, especially, in spinach. Monitoring the nitrate level also gives us a chance to have vitamin C-rich vegetables.

REFERENCES

Abraham-gutiérrez, J.H., Gil-muñoz, A., Sandoval-castro, E., Peña-olvera, B.V. and Almeida-acosta, F.H.E. (2004). Influence of bloodmeal and fertilizers on physical characteristics and yield of yam bean. *Terra* **22:** 475-483.

Ahrens, E., Elsaidy, S., Samaras, I., Samaras, F. and Wistinghausen, E. (1983). Significance of fertilisation for the post-harvest condition of vegetables, especially spinach. Environmentally Sound Agriculture-4th IFOAM Conference-Selected Papers (Cambridge MA), Lockeretz, W. (Ed.) Praeger, New York, pp. 229-246.

Anjana, Umar, S. and Iqbal, M. (2007). Nitrate accumulation in plants, factors affecting the process, and human health implications. *A review Agron. Sustain. Dev.* **27:** 45-57.

Anjana, Umar, S., Iqbal, M. and Abrol, Y.P. (2007). Are nitrate concentrations in leafy vegetables within the safe limits? *Curr. Sci.* **92:** 355-360.

Anonymous (2005). Commission Regulation (EC) No 1822/2005 of 8 November 2005 Amending Regulation (EC) No 466/2001 as regards nitrate in certain vegetables. Official Journal of the European Union L 293/11.

Anonymous (2006). Organic farming food quality and human health. A Review of the Evidence.

Anonymous (2007). Definition of Organic Agriculture. IFOAM. http://www.ifoam.org/growing_organic/definitions/doa/index.html.

Anonymous Anonymous (2013). FAO Statically Yearbook.

Anonymous (2008). Nitrate in Vegetables Scientific Opinion of the Panel on Contaminants in the Food Chain. European Food Safety Authority. The EFSA J. p. 689.

Baker, A. (1975). Organic vs inorganic nutrition and horticultural crop quality. *Hortic. Sci.* **10:** 50-53.

Besirli, G., Surmeli, N., Sonmez, I., Kasim, M.U., Basay, S. and Pezikoglu, F. et al. (2004). Organik Olarak Yetistirilen Ispanakta Verim, Kalite Özellikleri ve Nitrat Iceriklerinin Belirlenmesi. V. Sebze Tarimi Sempozyumu Bildiriler. Çanakkale Onsekiz Mart Üniversitesi Ziraat Fakültesi. Çanakkale.

Bhattacharjee, S., Dasgupta, P., Paul, A.R., Ghosal, S., Padhi, K.K. and Pandey, L.P. (1998). Mineral element composition of spinach. *J. Food Agric.* **77:** 456-458.

Citak, S. and Sonmez, S. (2010). Effects of conventional and organic fertilization on spinach (*Spinacea oleracea* L.) growth, yield, vitamin C and nitrate concentration during two successive seasons. *Sci. Hortic.* **126:** 415-420.

Citak, S and Sonmez, S. (2011). Monitoring of plant growth, yield, vitamin C and nitrate concentrations in organically and conventionally grown white head cabbage (Brassica oleracea var. capitata) in two successive seasons J. Of Food Agric & Envir. **9 (3-4)** pp: **501-505**.

Demir, K., Yanmaz, R., Ozcoban, M. and Kutuk, A.C. (1996). Ispanakta Farkli Organik Gübrelerin Verimlilik ve Nitrat Birikimi Üzerine Etkileri. GAP I. Sebze Tarimi Sempozyumu. Sanliurfa pp. 256-262.

Fytianos, K. and Zarogiannis, P. (1999). Nitrate and nitrite accumulation in fresh vegetables from Greece. *Bull. Environ. Contam. Toxicol.* **62:** 187-192.

Fujiwara, T., Kumakura, H., Ohta, S., Yoshida, Y. and Kameno, T. (2005). Seasonal variation of l-ascorbic acid and nitrate content of commercially available spinach. *Hortic. Res.* (Japan) **4:** 347-352.

Hornick, S.B. and Parr, J.F. (1990). Effect of fertiliser practices on the nutritional quality of crops. Agricultural alternatives and nutritional self-sufficiency for a sustainable agricultural system that respects man and his environment. The IFOAM International Scientific Conference (Ouagadougou, 2–5 January 1989), Witzenhausen, Ekopan pp. 244-254.

Ismail, A. and Fun, C.S. (2003). Determination of vitamin C, β-carotene and riboflavin contents in five green vegetables grown organically and conventionally. *Mal. J. Nutr.* **9:** 31-39.

Jablonska-Ceglarek, R. and Rosa, R. (2003). Forecrop green manures and the size and quality of white head cabbage yield. *Elect. J. Pol. Agric. Univ.* 6: Series Horticulture.

Kopp, H. (1992). Research on Organically versus Conventionally Grown Vegetables. Organic Agriculture, a Key to a Sound Development and Sustainable Environment-Proceedings of the 9th IFOAM Conference. Kopke, U. and Schultz, D.G. (Ed). Sao Paulo, Brazil, 16 Nov. 1992. Tholey-Theley, Germany pp. 324-330.

L'Hindorel, J.L. (2001). Nitrate and Man: Toxic, Harmless or Benefical? CABI Publishing.

Lairon, D., Termine, E., Gautier, S., Trouilloud, M., Lafont, H. and Hauton, J. (1986). Effects of organic and mineral fertilisations on the contents of vegetables in minerals, vitamin C and nitrate. Vogtmann, H., Boehncke, E. and Fricke, I. (Eds.). The Importance of Biological Agriculture in a World of Diminishing Resources-Proceedings of the 5th IFOAM International Scientific Conference Verlagsgruppe Witzenhausen pp. 249-260.

Lampkin, N. (2001). Organic farming. Old Pond Publishing.

Leuzzi, U., Cimino, G. and Dugo, G. (1996). Determination by means of Ionic chromatography of the anions present in vegetables produced organically. Rivista di Scienza dell"Alimentzione 25 **84:** 349-355.

Majumdar, D. (2003). The Blue Baby Syndrome. Nitrate Poisoning in Humans. Resonance, October.

Mirvish, S.S., Wallcave, L., Eagen, M. and Shubik, P. (1972). Ascorbate-nitrite reaction: possible means of blocking the formation of carcinogenic N-nitroso compounds. *Science* **177:** 65-68.

Moramoto, J. (2000). Comparison of nitrate content in leafy vegetables from organic and conventional farms in California. Organic Farming Research Foundation. Summer 2000, **8:** 23-25.

Rembialkowska, E. (2000). The Nutritive and Sensory Quality of Carrots and White Head Cabbage from Organic and Conventional Farms. The World Grows Organic: Proceedings from the 13th IFOAM Scientific Conference, Basel, Switzerland, **28-31:** 297.

Rembialkowska, E. (2007). The nutritive and sensory quality of carrots and white cabbage from organic and conventional farms. IFOAM 2000: The World Grows Organic. Proceedings 13th International IFOAM Scientific Conference, Basel, Switzerland, 20 to 31 August, 2000: 297 2000.

Rosen, C.J. and Allan, D.L. (2007). Exploring the benefits of organic nutrient sources for crop production and soil quality. *Hortic. Technol.*, October-December **17(4).**

Santamaria, P. (2006). Nitrate in vegetables: toxicity, content, intake and EC regulation. *J. Sci. Food Agric.* **86:** 10-17.

Schudel, P., Eichenberger, M., Augstburger, F., Klay, R. and Vogtmann, H. (1979). The influence of compost or NPK fertilisation on yield, vitamin C and nitrate content of spinach and spinach beet. *Schweizerische Landwirtschaftliche Forschung* **18:** 337-349.

Schultz, D.G., Zedow, E. and Kopke, U. (1992). Determining the quality of organic produce: extended quality parameters and quality index. Kopke, U. and Schultz, D.G. (Eds.). Organic Agriculture, a Key to a Sound Development and Sustainable Environment-Proceedings of the 9th IFOAM Conference. Sao Paulo, Brazil, 16 Nov 1992. Tholey-Theley, Germany pp. 338-348.

Scuphan, W. (1974). Nutritional value of crops as influenced by organic and inorganic fertiliser treatment. *Qualitas Plant-Plant Foods for Human Nutr.* **23:** 333-358.

Soel, (2011). FIBL Survey. Organic Agriculture Worldwide. www.soel.de

Stopes, C., Woodward, L., Forde, G. and Vogtmann, H. (1989). Effects of composted FYM and compound fertilizer on the yield and nitrate accumulation in three summer lettuce cultivars grown in an organic system. *Agric. Ecosys. Env.* : 555-559.

Termine, E., Lairon, D., Taupier-letage, B., Gautier, S., Lafont, R. and Lafont, H. (1987). Yield and content in nitrates, minerals and ascorbic acid of leeks and turnips grown under mineral or organic nitrogen fertilizations. *Plant Foods Human Nutr.* **37:** 321-332.

Wang, Z.H., Li, S.X. and Malhi, S. (2008). Effects of fertilization and other agronomic measures on the nutritional quality of crops. *J. Sci. Food Agric.* **88:** 7-23.

Warman, P.R. and Havard, K.A. (1997). Yield, vitamin and mineral content of organically and conventionally grown carrots and cabbage. *Agric. Ecosys. Env.* **61:** 155-162.

Zahradnik, A. and Petrikova, K. (2007). Effect of alternative organic fertilizers on the nutritional value and yield of head cabbage. *Hortic. Sci. (Prague)* **34:** 65-71.

6

Use of Organic Fertilizers for Crop Nutrition: Impact on Environment and Threats to Human Health

Zahoor Ahmad[1], Mitsuhiro Inoue[2] and Sadahiro Yamamoto[3]

[1]Department of Agriculture, University of Haripur, Haripur, Pakistan
[2]Arid Land Research Center, Tottori University, Japan
[3]Faculty of Agriculture, Tottori University, Tottori, Japan
e-mail: zahoor112@hotmail.com

ABSTRACT

Hundred and thousands of tons of bio-waste is generated every year across the glob. Most of the generated bio-waste makes its way to agricultural land as organic fertilizer. In addition to this organic fertilizer, more than 165 Mt of chemical fertilizers are added into the soil. Continuous use of these fertilizer has created sever environmental problems and threaten the human health. Chief among the environmental problems are eutrophication and emission of green house gasses (GHGs). Emissions of the GHGs covered by the Kyoto Protocol increased by about 70% (from 28.7 to 49.0 gigatonnes (Gt) CO_2-eq) from 1970–2004. For 2030, GHGs emission projection consistently show a 25–90% increase compared to 2000. Among GHGs, carbon dioxide (CO_2) is the most important anthropogenic GHG and its annual emissions have grown between 1970 and 2004 by about 80%, from 21 to 38 Gt, and represented 77% of total anthropogenic GHG emissions in 2004. Similarly, there are more than 150 pathogens present in different organic waste which may post threats to human health when applied on agricultural lands. The current chapter discusses some of these important issues taking into consideration especially the use of organic fertilizers for crop nutrition and the assessment of its potential impact on environment and threats to human health in detail.

Keywords: Organic fertilizers, Crop nutrition, Environment, pathogens, Human health

INTRODUCTION

The roots of farming began in the areas of the present-day Turkey and the Middle East about 10,000 years ago. Farming changed very little from early times until about 1700. In 1700s, an agriculture revolution took place which led to an increase in the production of crops. This increase came about in a large part by the final destruction of medieval institutions and the more general adoption of techniques and crops which had been known for a long time (Grigg 1974). Included in some of these changes was also the adoption of crops from the "new world" such as corn and potatoes, which produced a very large yield. In the 1850s, the industrial revolution spilled over to the farm with new mechanized methods that increased the production rates. The large changes were in the use of new farm implements. These new implements, combined with crop rotation, manure, and better soil preparation, led to a steady increase in the crop yield in Europe (Grigg 1974). Then came the most recent revolution called "Green Revolution". The Green Revolution began in 1943 when the Rockefeller Foundation and the Mexican government established the Cooperative Wheat Research and Production Programme to improve the agricultural output of the country's farms. In 1943, Mexico imported half of its wheat; in 1956, the Green Revolution had made Mexico self-sufficient; by 1964, Mexico exported half a million tons of wheat. That success was quickly adopted in India and Pakistan where, on an average, the programme tripled the harvest of wheat and other food crops per hectare.

It took nearly all of the human history from the first days of man on earth until the early 1800s to reach a global population of 1 billion. In just 200 years, we have managed to reach 6.5 billion. That means the population has grown more since 1950 than in the previous four million years. We are adding roughly 74 million people to the planet every year, a scary figure that will probably continue to increase. All these mouths need to be fed. All these people need clean water and a place to sleep. Green Revolution and spread of new cultivation techniques have played an important role to meet the fiber and dietary demands of rapidly growing world population. The projects within the Green Revolution spread technologies that had already existed, but had not been widely used outside of industrialized nations. These technologies included pesticides, irrigation projects, synthetic nitrogen fertilizer and improved crop varieties developed through the conventional breeding methods available at the time. After about six decades, the Green Revolution technologies are still going strong in terms of agricultural production, but we are discovering that there was a price to pay in terms of, among other things, environmental degradation. Use of any technology by itself is not harmful but the way it is used makes the difference and among the green revolution technology use of fertilizers (organic and inorganic) has important implications on environment and ultimately on human health.

FERTILIZERS IN AGRICULTURE

A fertilizer is a material that furnishes one or more of the chemical elements necessary for the proper development and growth of plants. Fertilizers can be divided into two main classes—inorganic fertilizers (also referred as chemical or mineral fertilizers) and organic fertilizers (various manures and plant residues). Like human beings, crop plants also need nutrients, water and air for survival. To meet their nutritional requirements plants solely depend upon the nutrients present in the soil. Until recent times, the growth of plants and animals, and hence the productivity of agriculture, was limited by the lack of nutrients since only small amounts are released annually from the rocks and soil minerals by weathering. With the passage of time, this realization that crops should be fed by supplementary nutrients led to a modern agriculture.

Organic Fertilizers

Chemically, a compound or a molecule containing carbon bounded to hydrogen is called organic. Organic compounds make up all living matter. The term "organic" is frequently used to distinguish "natural" products or processes from man-made "synthetic" ones. Natural fertilizers include manures or rock phosphate, as opposed to fertilizers synthesized from chemical feedstock. The word manure came from Middle English "manuren" meaning to till or cultivate land and initially from French "manoeuvre" meaning manual work alluding to the work which involved manuring land. There are two main classes of manures in soil management: green manures and animal manures. Compost is distinguished from manure in that it is the decomposed remnant of organic materials (which may, nevertheless, include manure). Mostly, animal manure is feces/excrement (variously called "droppings" or "crap", etc.) of plant-eating mammals (herbivores) and poultry, or plant material (often straw) which has been used as a bedding for animals, and thus is heavily contaminated with their feces and urine. Green manures are crops grown for the express purpose of plowing them under. In doing so, fertility is increased through the nutrients and organic matter that are returned to the soil. Leguminous crops, such as clover, also "fix" nitrogen through rhizobia bacteria in specialized nodes in the root structure. Other types of plant matter used as manure or fertilizer include the contents of the rumens of slaughtered ruminants and spent hops left over from making beer.

Before the introduction of mineral fertilizers in the nineteenth century, soil fertility was maintained mostly by the recycling of organic materials and crop rotations, which included nitrogen-fixing leguminous crops. From the very beginning of sedentary agriculture, farmers have recognized the basic problem of nutrient depletion. In all continents, slash burn techniques have been used to replenish nutrients that have been depleted by crops. Slash and burn is still practiced in many developing countries and relies on the slow, natural process of nutrient release from soil organic matter and soil minerals and on biological nitrogen fixation to accumulate nutrients in the form of biomass (UNIDO and IFDC 1998). The quantity of plant nutrients available in manures is insufficient to meet crop requirements. It is estimated that animal manures provide about 11 percent of the total N required for global food production (Smil 1999). The nutrient content of manures varies considerably and, if poorly stored, most of the N may be lost, resulting in the pollution of atmosphere. Compared with chemical fertilizers, the nutrient content of manures is low and variable, but they have the advantage of contributing to soil fertility. The organic matter improves the soil structure, reduces soil erosion, helps to store moisture, and provides fixation sites for certain plant nutrients. In many developing countries, a substantial proportion of the available manure is used for purposes other than application in agriculture, e.g., as fuel. In poor areas, people often use agricultural and livestock residues as a source of energy, and only a small proportion is returned to the land. Manure is bulky, odorous, difficult and expensive to store, handle and transport. Therefore, the ease of fertilizer use has sometimes led to the neglect of the use of organic materials in agriculture. However, in general, farmers apply manures and composts where they are available and where there is adequate labor to collect, store and spread them.

World bio-wastes

To meet the demands of increasing population, technological progress has made it possible to achieve remarkable improvements in land productivity, increasing per capita food availability, despite a consistent decline in per capita agricultural land. Meat demand in developing countries rose from 11 to 24 kg capita^{-1} yr^{-1} during the period 1967-1997, achieving an annual growth rate of more than 5% by the end of that period. Rosegrant et al. (2001) forecasted a further increase of 57% in global meat demand by 2020, mostly in South and Southeast Asia, and Sub-Saharan Africa. According to FAOSTAT (FAO 2008a), globally, approximately 56 billion land animals are reared and slaughtered for human consumption annually, and livestock inventories are expected to double by 2050, with most increases occurring in the developing world (Steinfeld et al. 2006). Similarly, the total production of poultry meat grew by about 75% during 2001-03 compared to 1991-93 (Steinfeld et al. 2006) and the demand of poultry is expected to increase by 83% by 2020 (Roy et al. 2002). According to FAO (FAOSTAT 2006), total production of meat and milk in Asian developing countries increased more than 12 times and 4 times, respectively, from 1961 to 2004. Since the per capita consumption of meat and milk is still much lower in these countries than in the developed ones, increasing trends are expected to continue for a relatively long time.

Although the poultry and livestock industry is playing an important role in meeting the dietary requirements of masses, at the same time this industry is producing huge amount of waste every year. As the poultry and livestock industry is expanding, the amount of waste generated each year has also increased tremendously. According to one estimate, cattle, hog, sheep and chook produce 14600, 1080, 600 and 275 kg waste per year per head (Miao and Zhao 2002). In total, livestock alone produces more than thirteen billion tons of waste each year, which is about 39% of the global waste. Similarly, in poultry industry, only in

United States, approximately 8.5 billion broiler chickens are produced annually (U.S. Department of Agriculture 2004). Every chicken produces between 1.46 and 2.67 kg of waste in its lifespan (Miner et al. 2000; Sharpe et al. 2004), resulting in an annual total of between 12 and 23 billion kg waste. Directly or indirectly, man himself is the biggest polluter of this planet. World population of 1998 generated about 5.92 billion solid waste (Marxsen 2001). Only in U.S., municipal solid waste is generated at an annual rate of 720 kg per person and totals to 194.5 million metric tons for the entire US population of 270.2 million people (Marxsen 2001). Similarly, according to one estimate, every person produces 55 g of dry sludge per day (Bradley et al. 1992). Multiplying times a global population of 6 billion people and annualizing the result indicates that 120,450,000 metric tons of sludge would be available from the entire world's population (Marxsen 2001). When activated sludge is co-settled with primary sludge, the total rises to 74 g per person per day. Adding the final tertiary sludge brings the per capita total up to 76 g per day, dry weight (Bradley et al. 1992). As an annual total for 6 billion people, this gives 166,440 thousand metric tons (Marxsen 2001). Adding up the municipal solid waste, sewer sludge, construction and demolition waste, industrial waste, and the three different kinds of agricultural and forestry waste (animal excrement, plant waste, and forestry harvesting residues) gives a surprisingly large total of 35.5 billion metric tons of annual global waste, as shown in Table 6.1 (Marxsen 2001).

Table 6.1: Estimate of global waste (Marxsen 2001)

Waste type	10^9 metric tons
Municipal solid waste	2.00
Sewer sludge	0.17
Construction and demolition waste	1.42
Industrial waste	1.42
Animal excrement	13.93
Agricultural plant waste	4.64
Forestry harvesting residues	11.94
Total	35.5

A huge amount of waste is produced every year, however, it does not imply that all waste generated is applied to agricultural lands. Most of the waste is recycled, incinerated, used as fuel (in developing countries), or goes to landfills. For instance, in USA, 54% of the municipal sludge is applied to land as fertilizer, 28% goes to landfills and 17% is incinerated (Marxsen 2001). Similarly, during 1970s at global level, 48% of the nutrients used by crops were derived from the soil, 13% from manure and 39% from fertilizers, as shown in Fig. 6.1 (Bockman et al. 1990; reported by Ayoub 1999). In 1990, the percentages changed to 30, 10 and 60, respectively. By 2020, the projected percentages will be 21, 9 and 70, respectively. This data predicted a decrease in the manure use in future, but there is a general realization among the scientific community that the use of manures will increase mainly due to the increase in livestock and poultry industry, as discussed earlier.

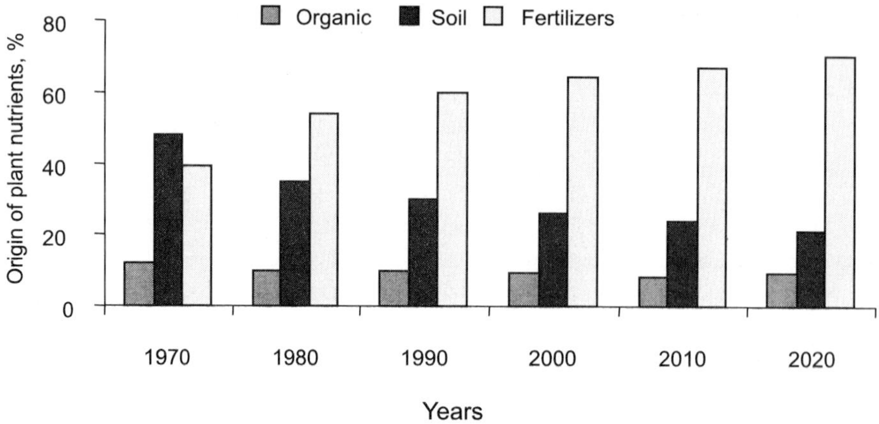

Fig. 6.1. Actual and projected origins of global plant nutrients (Bockman et al. 1990).

Inorganic Fertilizers

The modern fertilizer industry started little more than 150 years ago and was first envisaged by the German chemist Justus Von Liebig (1803-73). He set down his principles of plant nutrition and production in "Chemistry and Its Applications in Agriculture and Physiology", where he stressed the value of mineral elements derived from the soil in plant nutrition and the necessity of replacing those elements to maintain soil fertility. Now the role of mineral fertilizers in support of growing demand for agricultural commodities is well established (Oram and Hojjati 1995). Over the last 30 years, a positive correlation has been observed between cereal production and fertilizer use in developing countries, which currently use the bulk of mineral fertilizers and exhibit the largest rates of growth relative to developed countries (Figs. 6.2 and 6.3). Until the early 1990s, developed countries (excluding

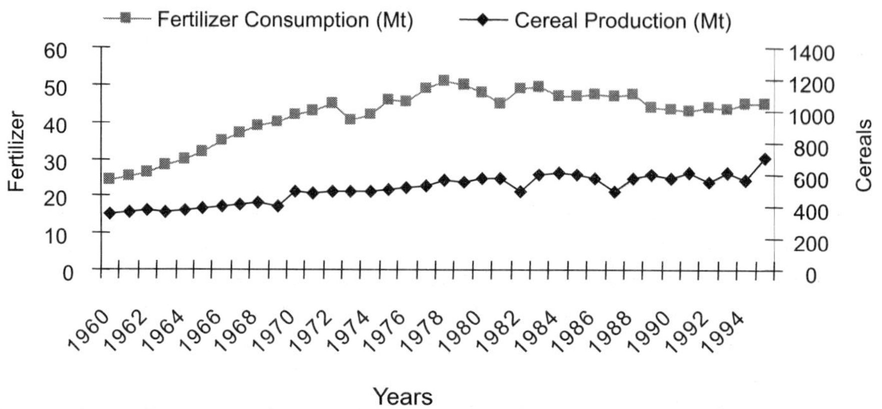

Fig. 6.2. Fertilizer consumption and cereal production in developed countries excluding Eastern Europe and FSU (Source: FAO 2000).

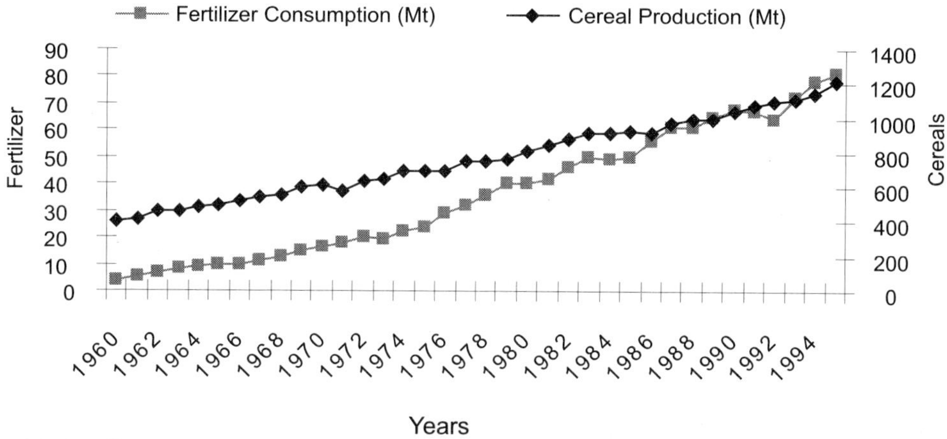

Fig. 6.3. Fertilizer consumption and cereal production in developed countries (Source: FAO 2000).

Eastern Europe and the Former Soviet Union) exhibited a similar correlation between cereal production and fertilizer use. More efficient use of fertilizer through improved timing, split applications, site-specific management, crop rotation, soil testing, etc. can facilitate higher yields with the same, or even less, fertilizer. Developed countries, especially those in North America and West Europe, appear to be exhibiting growing nutrient use efficiency, especially for nitrogen (Frink et al. 1998). As compared to the developed countries, the fertilizer use increased from 2.7 Mt in 1959/60 to 62.3 Mt in 1989/90 in developing countries with an annual growth rate of 10.5% (Table 6.2). Total world nutrient consumption rose to 161.2 Mt in 2006/07 (Patrick and Prud'homme 2008).

Table 6.2: Fertilizer use during 1959/60, 1989/90

Country category	Fertilizer Use (Mt nutrients)			Annual growth rate (%)
	1959/60	1989/90	2006/07[1]	1960-90
Developed	24.7	81.3	–	4
Developing	2.7	62.3	–	10.5
World	27.4	143.6	161.2	5.5

Source: Bumb and Baanate 1996; [1]Patrick and Prud'homme 2008

The demand and supply of chemical fertilizers depends largely upon the overall performance of agriculture. World agriculture (aggregate value of production) has been growing at the rates of 2.1–2.3 percent in the last four decades, with much of the growth originating in the developing countries (3.4–3.8 percent) (FAO 2007). The high growth rates of the latter reflected, among other things, developments in some large countries like India, Pakistan and China. Despite surging oil prices and an economic slowdown in the United States, the world economic context remained very robust in 2007. Reflecting continued rapid growth in Asia, world GDP in 2007 remained around

5% for the fourth consecutive year. However, after several years of extraordinary growth, the world economy is entering a depressed period. The situation deteriorated quickly during the third quarter of 2008. World GDP growth is seen as dropping from 5% in 2007 to 3.7% world GDP growth is seen as dropping from 5% in 2007 to 3.7% in 2008 (IMF 2008) and remained 2.71% during 2011, according to the International Monetary Fund (IMF) (IMF 2008). Fertilizer consumption rose sharply in 2007–08, boosted by strong agricultural commodity prices during the first half of 2008 and strong policy support in many Asian countries. After a 7.6% contraction in 2008/09, global fertilizer consumption started to recover (+5.4%) in 2009/10, reaching 163.9 Mt. Global fertilizer demand in 2011/12 is seen as rising more modestly, by 2.5%, to 176.4 Mt. Potassium demand is expected up by +5.5% while growth rates for N and P demand are seen as more moderate: +1.8% for N and +2.1% for P.

Over the long run, several studies have suggested that the expected yield and production will increase during the next 30 years, which will result increase in the use of fertilizers. For example, based on an econometric model, Bumb and Baanaante (1996) estimated that effective world demand for fertilizers would increase 1.2 percent annually between 1990 and 2020. Other less exhaustive studies have also ventured the estimates of future fertilizer use. Plucknett (1995) suggested that intensification of agriculture might require as much as 7 percent increase per year in fertilizer use to meet food needs. Alexandratos (1998) forecasted annual fertilizer growth rates of 3.8 percent between 1989–90 and 2010. Dyson (1998) estimated that until 2025, global use of mineral nitrogen will double, which implies annual compound rate of growth of 2.26 percent. Frink et al. (1999) expected nitrogen use to grow by 1.1 percent annually from 1990 to 2070 by using a deterministic model based on the population growth, gross domestic product and crop production. Most recently, one of the FAO reports estimated that world fertilizer consumption will rise well over 2% per year between 2008 and 2012, equivalent to an increment of 19.3 million fertilizer nutrient tones. While, the total world nutrient supply is projected to increase by 4.3 percent in the same period (FAO 2008b).

IMPACT OF FERTILIZERS ON ENVIRONMENT

Biophysically, the physical and biological factors along with their chemical interactions that affect an organism constitute the environment. The change in environment is a natural process, but human activities have accelerated the naturally occurring changes. Naturally occurring processes continually alter the Earth's atmosphere, topography, biomass and biofaunal loads and their distributions around the world. When these processes adversely impact the environment relative to the perceived needs of humanity, they are considered natural environmental problems. Remediation can be performed, but preventive measures seem difficult to provide. On the other hand, anthropogenic processes are those produced by man's activities in exploiting and modifying the environment. In general, there is a negative effect on the environment caused by such anthropogenic processes. In some cases, remediation can be performed and, moreover, preventive measures can also be provided. Increase in population and man's efforts to explore the secret of this universe and using these secrets to develop new technologies for its comfort has stimulated the environmental changes. Population has been theorized to be a key driving force of environmental change along with economic activities, technology, political and economic institutions, attitudes and beliefs (Ayoub 1999). Recent research evidence, however, demonstrates that population and affluence (per capita GDP) are primary determinants of a wide variety of environmental impacts (Ayoub 1999).

Soil is the only finite resource to feed the increasing population and soil fertility is vital to sustain the food productivity of soil. Fertilizers are required for sustained food production, but their widespread use has roused concern about the resulting environmental pollution. Hundred and thousand of tons of chemical and organic fertilizers are produced globally each year for application on agricultural lands, but only about half of the applied fertilizers only about half of the applied fertilizer is taken up by the crop. (Bockman et al. 1990), there is potential for marked economic losses and for negative environmental impacts. Chief among these problems are the excessive concentration of nitrogen compounds in the atmosphere (greenhouse gases), contribution of phosphate and nitrogen compounds to the eutrophication of surface waters, and risk to human health from manure-associated pathogens.

Greenhouse Gases

As early as 1896, Svante Arrhenius recognized that the Earth's atmosphere acts like a glasshouse, allowing passage of short wavelength solar radiation into the biosphere, trapping longer wavelength infrared radiation. There are many chemical compounds in the Earth's atmosphere that act as "greenhouse gases." These gases allow the sunlight to enter the atmosphere freely. When the sunlight strikes the Earth's surface, some of it is re-radiated back towards the space as infrared radiation (heat). Greenhouse gases (GHGs) absorb this infrared radiation and trap the heat in the atmosphere. Main greenhouse gases are carbon dioxide (CO_2), methane CH_4, nitrous oxide (N_2O), perfluorocarbons (PFCs), hydrofluorocarbons (HFCs), sulphur hexafluoride and ozone-depleting substances (ODS), Chlorofluorocarbons (CFCs), hydrochlorofluorocarbons (HCFCs), and halons.

Greenhouse gases are essential for maintaining the current temperature of the Earth; without them the planet would be so cold as to be uninhabitable (Karl and Trenberth 2003; Le et al. 2007). Natural sources of greenhouse gases include the Earth's ecosystem and some geological processes, while anthropogenic sources include industrial, transportation, residential, commercial and agricultural processes. Over time, if atmospheric concentrations of greenhouse gases remain relatively stable, the amount of energy sent from the sun to the Earth's surface should be about the same as the amount of energy radiated back into space, leaving the temperature of the Earth's surface roughly constant. In computer-based models, rising concentrations of greenhouse gases produce an increase in the average surface temperature of the Earth over time. Rising temperatures may, in turn, produce changes in precipitation patterns, storm severity, and sea level commonly referred to as "climate change."

Since preindustrial times, increasing emissions of GHGs due to human activities have led to a marked increase in atmospheric concentrations of the long-lived GHG gases. The predominant sources of the increase in GHGs are the combustion of fossil fuels. Atmospheric CO_2 concentrations have increased by almost 100 ppm in comparison to its preindustrial levels, reaching 379 ppm in 2005, with mean annual growth rates in the 2000–2005 period that were higher than those in the 1990s (Rogner et al. 2007). Emissions of the GHGs covered by the Kyoto Protocol increased by about 70% (from 28.7 to 49.0 gigatons (Gt) CO_2-eq) from 1970–2004 (by 24% from 1990-2004) (Barker et al. 2007). For 2030, GHG emission projections consistently show a 25–90% increase compared to 2000. Among GHGs, carbon dioxide (CO_2) is the most important anthropogenic GHG. Its annual emissions have grown between 1970 and 2004 by about 80%, from 21 to 38 Gt, and represented 77% of total anthropogenic GHG emissions in 2004 (Rogner et al. 2007). The rate of growth of CO_2-eq emissions was much higher during the recent 10 years period of 1995–2004 (0.92 Gt CO_2-eq yr^{-1}) than during the previous period of 1970-1994 (0.43 GtCO_2–eq yr^{-1}). Although occurring in smaller quantities, nitrous oxide (N_2O) is 270 times as active as CO_2 as a greenhouse gas. Most of the atmospheric NO, and NO_2 are recycled to soil and water bodies to form acid depositions. Nitric oxide (NO) and nitrogen dioxide (NO_2) have no absorption bands in the infrared part of the spectrum and have an atmospheric residence time of only a few days. They may, however, be involved in a number of atmospheric reactions, which affect the concentrations of other gases contributing to greenhouse warming. Both the gases catalyse ozone destruction in the atmosphere and are involved in the oxidation of CH_4 and CO, thus releasing the major greenhouse gas, CO_2. Methane (CH_4) is also an important culprit among GHGs. Its emissions rose by about 40% from 1970, with an 85% increase from the combustion and use of fossil fuels. Agriculture, however, is the largest source of CH_4 emissions (Barker et al. 2007).

Agriculture sector releases a significant amount of CO_2, CH_4, and N_2O in atmosphere (Paustian et al. 2004). Agriculture accounts for 10–12% of total global anthropogenic emissions of GHGs. This sector contributes about 47% and 58% of total anthropogenic emissions of CH_4 and N_2O (Smith 2007), respectively. Globally, agricultural CH_4 emission increased by 17% from 1990 to 2005, an average annual emission increase of 58 Mt CO_2–eq yr^{-1}. There is a wide range of uncertainty in the estimates of both the agricultural contribution and the anthropogenic total (Smith et al. 2007). Table 6.3 shows the contribution of agriculture to global GHGs and ammonia emissions. Achieving the increase in food production may require more use of N fertilizer, leading to possible increase in N_2O emissions, unless more efficient fertilization techniques and products can be found (Galloway 2003; Mosier 2002). Fertilizers contribute about 90% of all nitrous oxide (WRI/UNEP/ UNDP/WB, 1996), about 2% of the annual additions of CO_2 to the atmosphere of about 8.5 billion tons, and global ammonia emission of about 4.5 million tons (UNEP/UNIDO/IFA 1996). Agricultural N_2O emissions are projected to increase up to 35–60% by 2030 due to increased nitrogen fertilizer use and increased animal manure production (FAO 2003).

Table 6.3: The contribution of agriculture global greenhouse gas and ammonia emissions (adopted from Norse 2003)

Gas	Carbon dioxide	Methane	Nitrous oxide	Nitric oxides	Ammonia
Main effects	Climate change	Climate change	Climate change	Acidification	Acidification, Eutrophication
Agricultural sources	Land use change, especially deforestation	Ruminants (15)	Livestock (incl. manure applied to farmland) (17)	Biomass burning (13)	Livestock (incl. manure applied to farmland) (44)
(Estimated % contribution to total global emission)		Rice production (11)	Mineral fertilizers (8)	Manure and mineral fertilizers (2)	Mineral fertilizers (17)
		Biomass burning (7)	Biomass burning (13)		Biomass burning (11)
Agricultural emissions as % of total anthropogenic sources	15	49	66	27	93
Expected changes in agricultural emissions to 2030	Stable or declining	From rice; stable or declining From livestock; rising by 60%	35–60% increase		From livestock; rising by 60%

The role of organic fertilizer is more critical in this regard. Carbon dioxide is released largely from microbial decay or burning of plant litter and soil organic matter (Smith 2007). Methane is produced when organic materials decompose in oxygen-deprived conditions, notably from fermentative digestion by ruminant livestock, from stored manures, and from rice grown under flooded conditions (Mosier et al. 1998). Future CH_4 emission is expected to increase further due to the increase in livestock production. US-EPA (2006) forecasts an increase of 153% and 86% in emissions from enteric fermentation and manure management, respectively, from 1990 to 2020. Similarly, the area of rice grown globally is forecasted to increase up to 4.5% by 2030 (FAO 2003). In this scenario, it is forecasted that combined methane emissions from enteric fermentation and manure management will increase by 21% between 2005 and 2020 (US-EPA 2006). N_2O is generated by the microbial transformation of nitrogen in soils and manures, and is often enhanced where available nitrogen (N) exceeds plant requirements, especially under wet conditions (Smith and Conen 2004).

Pollution by Nitrogen Fertilizers

Elemental N (N_2) constitutes 99.8% of global N, and 78% of the atmosphere is N_2. The gas in the atmosphere is the ultimate N source and sink, but it has low chemical reactivity under most natural conditions and only some bacteria can utilize it. Nitrogen is available to support the growth of plants and microbes only after it is converted into reactive forms (Galloway et al. 2003), such as ammonium (NH_4^+) and nitrate (NO_3^-).

Nitrogen exists in various forms:

a) Gaseous:
   i. nitrous oxide (N_2O)
   ii. nitric oxides, NOx a mixture of nitric oxide (NO) and nitrogen dioxide (NO_2)
   iii. ammonia (NH_3)

b) Ionic:
   i. nitrate (NO_3^-)
   ii. ammonium (NH_4^+)

c) Organic:
   i. urea [$CO(NH_2)_2$]
   ii. substances that contain N in living and dead organisms, e.g., proteins, enzymes, etc.
   iii. N in transformed organic residues of low reactivity, e.g., humus.

"Mineral N" is a collective term for NO_3^- and NH_4^+, and includes urea, a simple organic molecule which readily hydrolyses to produce NH_4^+. The principal forms of N in soil are ammonium (NH_4^+), nitrate (NO_3^-) and organic substances. Most of the soil N is contained in the organic matter. This is slowly converted by microbes to NH_4^+, and then other microbes convert NH_4^+ rapidly to NO_3^-, which is the main mineral N form in most soils. Nitrate and NH_4^+ are directly available to plants. Plants can also take up small amounts of gaseous NH_3 and soluble organic N compounds such as urea and amino acids. Nitrogen in other organic compounds is not available until these are decomposed. NH_4^+ can be held on negatively charged sites of clay minerals and organic (humic) compounds. This reduces its mobility in the soil compared to the more mobile NO_3^- ion. Some clays bind NH_4^+ so tightly that the availability to plants is severely restricted, also slowing the conversion of NH_4^+ to NO_3^-. Nitrate is one of the most mobile ions in the soil. Generally, soils have a higher binding capacity for cations than anions, but some anion-binding capacity exists. However, other anions (like phosphate and sulphate) are bound in preference to nitrate. Nitrate, in excess of what is taken up by plants, is therefore at risk to loss by leaching when excess rainfall or irrigation water drains through the soil.

During the past two centuries, and especially over the last five decades, humans have substantially altered the global nitrogen cycle (as well as the global cycles of other chemical elements), increasing both the availability and the mobility of nitrogen over large regions of Earth (Carpenter et al. 1998; Howarth et al. 2000; Galloway and Cowling 2002). Overall, fertilizer application, legume crops and fossil fuel burning now deposit approximately 140 Mt new fixed N annually on land

(Table 6.4). This exceeds the natural N inputs. In addition, N stored in soils is mobilized through tillage, burning vegetation, draining of wetlands and conversion of land from natural vegetation (grassland, forest) to agricultural use. The increased amount of N moving in the N cycle is at the core of environmental issues related to N in agriculture (Vitousek et al. 1997). Now, nitrate is a common surface water and groundwater contaminant that can cause health problems in infants and animals, as well as the eutrophication of water bodies (Fennesy and Cronk 1997).

Table 6.4: Estimates of global N fixation (Laegrid et al. 1999)

		N (Mt year^{-1})	
			Most probable value
Source	Range of estimates	Natural	Human activity
Biological fixation		100	40*
Land	90–140		
Ocean	1–120		
Fertilizers			79
Combustion of fossil fuel			22
Lightning	5–25	13	
Total		113	141

*Relates to the use of legumes in agriculture.

Nitrogen added to the soil is taken up by plants, becomes incorporated in the soil's organic matter, or is lost from the agricultural system. Nitrogen loss from the soil occurs through leaching, denitrification or through runoff water. As mentioned earlier, soils have low NO_3^- binding capacity, therefore, they follow the water movement. Usually leaching takes place when the soil is fully wet, and the input of water (by rain or melting snow) exceeds the water loss by evapotranspiration. The ease of water and NO_3^- movement in the soil depends on the soil structure and porosity (Gaines and Gaines 1994). As expected from the mechanism of leaching, land under fallow has a larger potential for NO_3^- leaching than the land covered by crops. The mineralization of soil organic matter generates NO_3^-, but fallow land has no plants that can utilize N and water, and in many regions the highest N losses occur during fallow. In contrast, grassland repeatedly cut and harvested has a dense and extensive root system and a very long season for growth and NO_3^- uptake. Consequently, NO_3^- losses by leaching from grasslands are usually small. However, grazing animals leave dung and urine patches on the pasture. These spots contain more N than the plants at the spot can utilize, and NO_3^- can leach from these spots even though the loss by leaching from the rest of the field is small. Hence, there is a greater risk of leaching in grazed than in mown grassland. A single spot can have an N concentration corresponding to more than 600 kg N ha^{-1} (Whitehead 1995).

Crops also have a major influence on water movement in soil. Plant roots take up water with dissolved nutrients and emit water to the atmosphere by transpiration from the leaves. Water is also lost by evaporation from the soil surface. This combined evapotranspiration often prevents leaching in the growing season, and can draw water from deeper layers to the surface. Such upward movement of water can make NO_3^- available, that has temporarily been out of reach, for plant uptake (Kuhlmann et al. 1989). A deep and extensive root system enables crops to utilize N efficiently and minimizes the risk of leaching. Crops differ in the extent of their root development as well as in the amount of N they need in order to produce a satisfactory yield. Some field vegetables have shallow roots and a high N demand, and some, e.g., cauliflower, contain only a small amount of N in the produce. For these crops, most of the applied N remains in the soil and in the plant residues left on the soil (Kuhlmann and Engels 1989). Hence, NO_3^- leaching from vegetable production can be substantial.

The easiest form of nitrogen loss is through denitrification. Large amounts of N_2 are emitted as the end product of denitrification. The environmental concerns related to the gaseous emissions of fixed N are:

i. Acidification of soils and surface waters;
ii. Eutrophication and loss of biodiversity in surface waters and remote land areas;
iii. Increased greenhouse gas concentrations in the atmosphere;
iv. Increased formation of near-ground ozone through reactions between NO_x and volatile organic carbon compounds.

Certain soil bacteria that thrive in saturated (anaerobic) soil conditions will convert nitrate-N to oxygen and nitrogen gases. Volatilization of the nitrogen gas can result in N losses of as much as 5% of the available nitrate-N per day (Robert 2004). Soils at greatest risk to denitrification N loss are those that are naturally heavy and poorly drained, plus fields with significant levels of soil compaction that restricts natural drainage. Urea-based nitrogen fertilizer products are susceptible to volatilization losses of nitrogen if surface-applied and not incorporated. Urease enzymes in the soil and plant residues convert the urea component to free ammonia gas. If this conversion occurs at the soil surface and is accompanied by warm sunny days, as much as 15~20% of the urea-based nitrogen may volatilize within a week after application (Bundy 2001). If a half-inch or more of rain occurs within the first 24 hours after surface application, the risk of subsequent volatilization also drops to essentially zero (Bundy 2001). If the urea-based product is injected or mechanically incorporated after application, the risk of volatilization is essentially zero. The risk of volatilization loss is greatest with high-residue cropping systems and surface soil pH levels greater than 7.0. Volatilization risk is also high in the soils with low buffer capacity.

Loss of nitrogen from the agricultural lands can also occur through runoff water and soil sediments. Surface runoff can remove large quantities of nutrients from the soil in both dissolved and sediment-bound forms (Gifford and Busby 1973; Lowrance and Williams 1988). Soluble nitrogen is transported by the surface flow, while insoluble forms of nutrients get adsorbed to sediment particles (mainly NH_4^+) are transported by erosion. The ultimate destiny of dissolved or particulate nitrogen is some lake, river or directly into sea. This process is very complex and is influenced by a number of factors like soil type (Merz and Bárdossy 1998), intensity and duration of precipitation, properties of the soil surface (Emmett, 1980), type of land cover (Liang et al. 2002), the change in land cover (Noah and Emmanuel 2002) and the type of organic and inorganic fertilizer applied (Zahoor et al. 2007). Nitrogen removed from the soil reaches streams, lakes, rivers or ends up in oceans. The transport of NO_3^- and NH_4^+ by rivers has been estimated to be about 20 Mt N yr^{-1} (Seitzinger and Kroeze 1998), and the total N transport (including organic N) is assumed to be in the range 40–80 Mt N yr^{-1}. However, only a fraction of the N and P entering rivers reaches the ocean (Behrendt 1996). A large part is removed from the water through sedimentation and, for N, also through denitrification. Biological denitrification in the sediments also contributes N_2O to the greenhouse gases.

Phosphorus and Water Pollution

Continuous long-term phosphorus applications can raise phosphorus levels above those required for optimum crop yields. Once the phosphorus levels become excessive, the potential for loss in runoff and drainage water is greater than any agronomic benefit of further applications. The phosphorus level in the surface soil determines the phosphorus loads in runoff and the proportions of soluble and particulate phosphorus lost. Increased residual levels in the soil lead to increased loadings in surface waters resulting in water pollution.

Water pollution involves the release into lakes, streams, rivers, and oceans of substances that become dissolved or suspended in the water or get deposited upon the bottom and accumulate to the extent that they interfere with the functioning of aquatic ecosystems. The use of chemical fertilizers, pesticides, or other materials on watershed lands is an additional factor contributing to water pollution. The runoff from septic tanks and the outflow of manures from livestock feedlots along the watershed are the sources of organic pollutants. Other important causes of surface water degradation are siltation caused by erosion from agriculture, logging and construction (which also contributes to eutrophication), acidification from atmospheric sources and mine drainage, contamination by toxins, introduction of exotic species, and hydrologic changes (NRC 1992). Similarly, the output of industries, agriculture, and urban communities generally exceeds the biologic capacities of aquatic systems, causing the waters to become choked with an excess of organic substances, and organisms to be poisoned by toxic materials. When organic matter exceeds the capacity of the microorganisms in water that break it down and recycle it, the excess nutrients in such matter encourages rapid growth, or blooms, of algae. When they die, the remains of the dead algae add further to the organic wastes already in the water and eventually, the water becomes deficient in oxygen. Anaerobic organisms (those that do not require oxygen to live) then attack the organic wastes, releasing gases such as methane and hydrogen sulfide, which are harmful to the oxygen-requiring (aerobic) forms of life. The result is a foul-smelling, waste-filled body of water, which is a growing problem in many freshwater lakes of the world. The process by which a lake or any other body of water changes from a clean, clear condition with a relatively low concentration of dissolved nutrients and a balanced aquatic community to a nutrient-rich, algae filled body and thence to an oxygen-deficient, waste-filled condition is known as accelerated eutrophication.

Nutrients are fundamental for plant growth in all ecosystems, as much in water as on land. An increase in the available nutrients or nutrient enrichment of a water body is called eutrophication (from the Greek word "eutrophos", meaning nourishing, well nourished, providing good nourishment). In 1907, C.A. Weber invented the terms (German adjectival) *eutrophe* (high nutrient contents), *oligotrophe* (low in nutrients, low algal growth and clear water) and *mesotrophe* (nutrient level and water quality is in between the eutrophic and oligotrophic lakes) to describe different states of the flora in German peat bogs, as the bogs changed with time through the accumulating material (Weber 1990). Simply, the term eutrophication describes the enrichment of freshwater with plant nutrients because of human activity. "Plant nutrients" mainly refer to inorganic compounds of nitrogen and phosphorus, essential for the growth of photosynthetic organisms. They also include dissolved silica, essential for the growth of diatoms, a class mostly consisting of free floating phytoplankton with silica skeletons (almost like tiny glass boxes), as well as micronutrients such as iron and manganese. Algae with a silicate skeleton (diatoms) provide food for organisms higher in the food chain, and silicon supply is necessary for these beneficial organisms. The algae utilize these three nutrients at an N: P: Si ratio of approximately 7:1:7 (or as N: P_2O_5: Si, 3:1:3) on a weight basis. A low N:P ratio in freshwater favors the production of N-fixing cyanobacteria (blue-green algae), while a high N:P ratio combined with low Si availability favors the production of dinoflagellates, the algal species forming 'red tides' that discolor the water and can be toxic.

Enrichment of nutrients most commonly results in the excessive stimulation of phytoplankton growth, but it may also trigger the growth of larger plants (macrophytes) on the lake or sea floor in shallow areas. Advanced eutrophication of surface water leads to problems with its use for fisheries, recreation, industry, or drinking due to the increased growth of undesirable algae and aquatic weeds and oxygen shortages caused by their senescence and decomposition. Also, many drinking water supplies throughout the world experience periodic massive surface blooms of cyanobacteria (Kotak et al. 1993). These blooms contribute to a wide range of water-related problems including summer fish kills, unpalatability of drinking water, and formation of trihalomethane during water chlorination (Palmstrom et al. 1988). Consumption of cyanobacterial blooms or water-soluble neuro- and hepatoxins released when these blooms die can kill livestock and may pose a serious health hazard to humans (Lawton and Codd 1991).

Eutrophication affects rivers, lakes and estuaries, as well as some coastal areas and enclosed water bodies. It can be either natural or caused by human activities. Though the definition is simple, the phenomenon, however, is a complex one because natural variations in the nutrient supply to the aquatic environment are very large. Eutrophication of most freshwater around the world is accelerated by P inputs (Schindler 1977; Sharpley et al. 1994). Although nitrogen (N) and carbon (C) are also essential to the growth of aquatic biota, most attention has focused on P inputs because of the difficulty in controlling the exchange of N and C between the atmosphere and water and the fixation of atmospheric N by some blue-green algae. Therefore, P is often the limiting element, and its control is of prime importance in reducing the accelerated eutrophication of freshwaters (Schindler 1977). Eutrophication, caused by excessive inputs of P and N, is a common and growing problem in lakes, rivers, estuaries, and coastal oceans in many parts of the world (Smith 1998). Freshwater eutrophication has been a growing problem for decades (NRC 1992). But, as in lakes and rivers, an increase in the supply of nutrients can have profound effects on receiving water quality in both estuarine and coastal marine ecosystems. Now, eutrophication is also widespread and rapidly expanding in estuaries and coastal seas of the developed world (Nixon 1995). The marine environment consists of estuarine and coastal waters and open seas. This is a more complex system than the confined freshwaters, and insight in marine eutrophication is less developed than for lakes and rivers. Eutrophication in the marine environment is restricted to some estuarine and coastal areas and partly enclosed water bodies, while the influence on the open ocean is negligible (Sampat 2000). In most temperate estuaries and coastal ecosystems, N is the element most limiting to primary production and most responsible for eutrophication. Although N is the major factor in eutrophication of most estuaries and coastal seas, P is also an essential element that contributes to coastal eutrophication. It is, in fact, the dominant control of primary production in some coastal ecosystems.

Trophic status	Definition, characteristics
Oligotrophic	Poorly nourished, clear waters with limited sediment and biological activity. Total P concentration < 10 ppb
Mesotrophic	Intermediate productivity and clarity. Total P concentration: 10 to 25 ppb
Eutrophic	High nutrient content. Potential water treatment problems with taste and odor. Total P concentration: 25 to 100 ppb
Hypereutrophic	Highly productive, murky waters. Can support cyanobacteria that may produce toxins. Total P concentration >100 ppb

From agricultural point of view, non-point source of water pollution is very important. There is a range of situations which might lead to increased rates of non-point nutrient transfer from agricultural land to water, some of which include: (a) increased contamination of storm run-off from farm buildings and livestock holding yards; (b) accumulated build-up of N and P in the soil profile; (c) changes in the storage, form and method of application of manures and fertilizers; (d) changes in land management/soil conditions that affect either storm run-off, or soil erosion rates; and (e) the extent of atmospheric exchanges of N between the lowlands and the uplands (Edwards and Withers 1998; Chapman 1999). Whilst landscape characteristics such as climate, topography and soil type govern the inherent risk of nutrient loss within a given locality, historic and current nutrient inputs and land management practices can exacerbate nutrient losses from different agricultural land uses. The significance of these factors varies appreciably between the two nutrients, and between different geographic regions and farming systems.

FERTILIZERS AND THREATS TO HUMAN HEALTH

The birthright of all living things is health. This law is true for soil, plants, animals and man: the health of these four is one connected chain. Any weakness or defect in the health of any earlier link in the chain is carried on to the next and succeeding links, until it reaches the last, namely, man.

Chemical and organic fertilizers play an important role in meeting the food demand of growing population. But, on the other hand, these fertilizers may pose threat to human health. Chemical fertilizers can contain harmful toxic metals that make the farms, lawns and gardens dumping grounds for toxic waste. Fertilizers intended for enriching the soil often contain numerous toxic metals that have no benefit for crop and plant growth. In US, California Public Interest Research Group (CALPIRG) Charitable Trust and Washington's Safe Food and fertilizer tested 29 fertilizers and reported that 22 fertilizers contained toxic heavy metals (Shaffer 2001). These metals are linked to either ecological or human health hazards. These results indicate that fertilizers often contain high levels of harmful toxic metals that exceed the levels of concern. The introduction of toxic substances like lead and cadmium to farm soils can adversely affect the growing conditions and result in increased toxic accumulation, as these metals are highly persistent in soils. This can negatively affect the critical growing requirements, such as soil acidity or the solubility of beneficial metals like zinc in the soils. Some crops are more likely than others to absorb non-nutrient toxic substances from the soils. For example, fruits and grains can absorb lead, and lettuce; corn and wheat can absorb cadmium from the soils (Wilson 1997). This means that our food supply is at risk of fertilizers contaminated with toxic substances that could threaten human health. Similarly, organic sources of fertilizers, mainly manure and biosolids (treated sewage sludge), can also endanger human health directly through germs or indirectly through nutrients and heavy metal contamination of soil and water resources.

All animals including pets, livestock, wildlife and humans are hosts to various microorganisms (germs), some of which may cause diseases. Many types of bacteria, parasites, protozoa and viruses are found in humans and animals, as well as in soil and water. Not all the germs from livestock are hazardous to humans and other animals; only a few strains have the potential to cause illness. Germs that make people ill are referred as pathogens. Just because a pathogen is present in the environment does not automatically mean that a person will contact a disease. In the same way, one sick animal does not mean that all of the animals in a herd or flock will become ill. For example, *Salmonella*, *Campylobacter* and *E. coli* O157 are common bacteria that can exist in the intestines of various animals, without the animals showing any signs of disease. These bacteria can also be present in manure and are potential causes of food/waterborne illness in humans.

The transmission of pathogens from animals to people is possible. The most common way is through direct contact such as touching animal, animal manure, animal products or drinking contaminated water. Livestock held in confined animal feeding operations serve as an important reservoir for significant human pathogens (zoonotic). Rodents, birds, insects, contaminated feed or equipment may also spread the pathogens. However, the main culprit is animal excreta, i.e., manure or slurry. Globally, agricultural (e.g., manure) and municipal (e.g., biosolids) wastes with potential uses in agriculture are generated in billions of tons annually, as discussed earlier. Many of these materials are spread, sprayed or otherwise applied to the agricultural land because of the benefits they provide. Composts are also used as a growth medium in home gardening, ornamental nurseries, and greenhouses (Freeman and Cawthon 1999).

A potential risk arising from the disposal of animal waste of fecal origin is the spread of enteric pathogens (Pell 1997). Different manures applied to the soil may turn into threats for human and animal health. More than 150 microbial pathogens have been identified from all the animal species that can be transmitted to humans by various routes (USDA 1992; USEPA 1998). Many of these can be found in animal feces as well as in the urine. Common zoonotic diseases transmitted by the farm animals and human wastes during the last decade are given in Table 6.5. Only in US there are approximately 76 million cases of food-borne illness annually, resulting in 325,000 hospitalizations and 5,000 deaths (Mead et al. 1999). Only an estimated 14 million of the cases are attributed to known pathogens, whereas 62 million illnesses are of unknown etiology (Doyle 2000). Similarly, in the United Kingdom, there are an estimated 1.34 million cases of gastroenteritis and food-borne illness annually (Adak et al. 2002).

Table 6.5: Recent examples where manure has been implicated as the source of pathogens (Smith and Perdek 2003; Guan and Holley 2003)

Location and date	Pathogen(s)	Human morbidity and mortality	Suspected source(s)
Somerset, UK 1997	*E. coli* O157:H7	8 cases	Contaminated mud at an open air music festival, Infected cattle (650 cows) grazed on the site
Scotland, UK 1999	*E. coli* O157:H7	6 cases	Contamination of untreated, unprotected private water source in a rural area where sheep and deer grazed freely
Washington county, USA 1999	*E. coli* O157:H7 & *Campylobacter* spp.	2 deaths, 116 cases	Runoff at fairgrounds
Ontario, Canada 2000	*E. coli* O157:H7 & *Campylobacter* spp.	6 deaths, 1346 cases	Pathogens from cattle manure on adjacent farms entered municipal well following heavy rains and flooding
Saskatchewan, Canada 2001	*Cryptosporidium parvum*	1907 cases	Surface river water was probably contaminated from some point upstream

Man is in contact with the soil permanently, either directly or indirectly via food, water and air, and thus soil may act as a vector and source of important human diseases. Pathogens can be transmitted from the animals to humans when manure used as a fertilizer for food crops is eaten raw, and by storm water runoff from manured surface to surface-waters or by its percolation to groundwater. Infection starts when the pathogens enter our mouth; in other words, when small (usually invisible) amounts of human or animal feces enter our mouth. Unfortunately, this happens more often than we would like to think about. The causes of illness include consumption of contaminated food, unpasteurized (raw) milk, water that has not been disinfected, ground beef, venison, sausages, dried (non-cooked) salami, cheese, orange juice or contact with the feces of infected people (Friedman et al. 1999). Ready-to-eat products such as unpasteurized apple cider and sprouts also pose threats to human health (Guan and Holley 2003). Apples used for making cider can be contaminated when dropped on ground fertilized with the manure. Pathogens on the skin of apples can spread throughout the batch of cider (Besser et al. 1993). Similarly, sprout seeds may initially become contaminated on the farm through the use of manure as fertilizer, and subsequently during the sprouting process, pathogens increase to high levels within the seed lot (Taormina et al. 1999). Root crops such as radish and carrots as well as leafy vegetables like alfalfa, radish, sprouts (Breuer et al. 2001) and lettuce, where the edible part touches the soil, also present a great risk. Despite the recognition of this hazard, there has been an increase in outbreaks associated with these products in recent years.

Table 6.6 shows different groups of pathogens present in the human wastes and animal manures, which may cause illness in humans and other animals. The main groups include bacteria, parasites, viruses and protozoa.

Table 6.6: Pathogens in human wastes (sewage and sewage sludge) and animal manures (Filip 1988; USEPA, 1999; Gerba and Smith 2005)

Pathogens	Disease
Bacteria	
Salmonella (1700 serotypes)	Salmonellosis (food poisoning), typhoid fever
Shigella spp.	Bacillary dysentery
Mycobacterium Tuberculosis	Tuberculosis
Vibrio cholerae	Cholera
Escherichia coli (pathogenic strains)	Gastroenteritis
Yersinia enterocolitica	Acute gastroenteritis (including diarrhea, abdominal pain)
Clostridium perfringens	Gastroenteritis, gangrene
Clostridium botulinum	Botulism
Listeria monocytogenes	Encephalitis
Campylobacter jejuni	Gastroenteritis
Virus	
Enterovirus	
Poliovirus	Poliomyelitis
Coxsackievirus	Meningitis, pneumonia, hepatitis, fever, cold-like symptoms, etc.
Echovirus	Meningitis, paralysis, encephalitis, fever,
Parvovirus	Cold-like symptoms, diarrhea, etc.
Hepatitis A virus	Infectious hepatitis
Human calicivirus	Epidemic gastroenteritis with severe diarrhea
Rotavirus	Acute gastroenteritis with severe diarrhea
Hepatitis E virus	Hepatitis
TT hepatitis	Hepatitis
Astroviruses	Epidemic gastroenteritis
Adenovirus	Respiratory infections
Protozoa	
Entamoeba histolytica	Amoebic dysentery
Giardia lamblia	Giardiasis (including diarrhea, abdominal cramps, weight loss)
Balantidium coli (rare)	Diarrhea and dysentery
Cryptosporidium parvum	Gastroenteritis, cryptosporidiosis (diarrhoeal illnesses)
Acanthamoeba (rare)	Meningoencephalitis
Toxoplasma gondii	Toxoplasmosis
Helminths Worms	
Ascaris lumbricoides	Digestive and nutritional disturbances, abdominal pain, vomiting, restlessness
Ascaris suum	May produce symptoms such as coughing, chest pain, and fever.
Ancylostoma spp.	Hookworm disease
Necator amaricanus	Hookworm disease

Table 6.6: Contd.

Pathogens	Disease
Enterobius vermicularis	Human pinworm
Strongyloides stercoralis	Small roundworm
Trichuris trichiura	Abdominal pain, diarrhea, anemia, weight loss
Taenia solium	Human tapeworm
Taenia saginata	Nervousness, insomnia, anorexia, abdominal pain, digestive disturbances
Toxocara canis	Fever, abdominal discomfort, muscle aches, neurological symptoms
Hymenolepis nana	Taeniasis, dwarf tapeworm
Fungi	
Candida spp.	Mycoses (skin and systemic)
Trichosporon cutaneum	Skin mycosis
Aspergillus fumigatus	Lung mycosis
Trichophyton spp.	Skin mycosis
Epidermophyton spp.	Skin mycosis
Microsporum spp.	Skin mycosis

Bacteria

Many pathogenic bacteria are present in human waste and animal manures. The bacterial pathogens most important with regard to human health include *Salmonella* spp., *Escherichia coli* O157:H7, *Campylobacter jejuni*, *Yersinia enterocolitica* and *C. perfringens*. *Salmonella* are *Enterobacteriaceae* which are widely distributed in the environment and include more than 2000 serotypes (Venglovsky et al. 2006). They are one of the most predominant pathogenic bacteria in wastewater and include serotypes that cause typhoid, paratyphoid fever and gastroenteritis. It has been estimated that 0.1% of the population excretes *Salmonella* at any given time (Venglovsky et al. 2006). This pathogen produces an endotoxin that causes fever, nausea and diarrhea and may be fatal if not treated properly by antibiotics (Bitton 1994). *Escherichia coli* O157:H7 (shortly called *E. coli* O157 or O157) is a major bacterial pathogen that has been implicated in diarrhea, hemorrhagic colitis, and hemolyticuremic syndrome. *E. coli* O157:H7 was first recognized as a food-borne pathogen in 1982 during an investigation into an outbreak of hemorrhagic colitis (bloody diarrhea) associated with the consumption of contaminated hamburgers (Riley et al. 1983). The following year, Shiga toxin, produced by then little-known *E. coli* O157:H7, was identified as the real culprit. Since 1982, *E. coli* emerged as a major cause of bloody and non-bloody diarrhea, causing thousands of infections with substantial illness and deaths each year (Mead and Griffin 1998). The U.S. Centers for Disease Control (CDC) estimated more than 70,000 cases of infection in humans each year by *E. coli* alone, 85% of them being food-borne in origin (Mead et al. 1999). Food-borne illness surveillance data from 1990–1998 in the USA showed that contaminated produce (fruit and vegetables including juices and salads) accounts for about 24% of the illness outbreaks (41% of cases), meats account for about 29% (20% of cases), and seafood about 14% (8% of cases) (Griffiths 2000).

Parasites

At present, helminthic pathogens play a minor role in industrialized countries (Venglovsky et al. 2006). Many helminths are, however, endemic in most tropical, less industrialized countries, and have, via waste-fed aquaculture, various potential transmission patterns. *Ascaris* (roundworms), the most common helminthic pathogen in developing countries, may be carried passively by fish and other aquacultural product, although its major foci are soils and waste-fertilized crops (Venglovsky et al. 2006). The three main transmission routes for *Ascaris* are: transmission from feces-contaminated surfaces and materials, transmission to workers in fields that have been fertilized with feces or sewage, and transmission by consumption of vegetables fertilized with feces or sewage (Feachem et al. 1983; Varadyova et al. 2001).

Viruses

More than 140 enteric viruses can be transmitted by biosolids (Venglovsky et al. 2006). The caliciviruses, adenoviruses, hepatitis A and E viruses, astroviruses, and rotaviruses are of particular concern. Viruses can easily adsorb on the soil surface, move with surface runoff, and can percolate to deeper soil layers to contaminate groundwater. Zoonotic viruses of humans continue to be discovered or appear to re-emerge as important human pathogens. Recent among these are the SARS virus (a coronavirus), monkeypox virus (a poxvirus), and hepatitis E virus (Venglovsky et al. 2006).

Protozoa

Protozoa are microscopic, single-celled organisms that belong to the kingdom Protista. Although they are single-celled, their structure is highly complex. About 66,000 species have been described, of which 10,000 are parasitic. Only a few species are important disease-causing parasites in humans. Protozoa of primary concern in watersheds are *Cryptosporidium parvum* and *Giardia* species. These protozoa cause mild to severe diarrhea and potentially shorten the lifespan of immune-compromised individuals, such as those with AIDS. Neither *Cryptosporidium* nor *Giardia* can reproduce outside the host. They also produce a water-borne cyst (Giardia) or oocyst (Cryptosporidium) that is resistant to chemical disinfectants (Sterling 1990).

As a number of bacteria, viruses and parasites reach the soil through contaminated manure, biosolid, slurry, water or wastewater, it is not quite certain that when a pathogen reaches the soil, it will also pose threat to the human health. Many factors influence the pathogens to become a threat for any life. The number and nature of such factors is so complex that it is quite difficult, even impossible, to predict exact environmental conditions that may favor or limit the survival of these organisms in the soil. One set of environmental condition may favor the living of one community of pathogens, while the same environment may become lethal for others. Certain pathogens exit the host in an environmentally resistant form (e.g., cyst) that can remain viable in the environment depending on the pathogen type and surrounding environment. This allows the organism an opportunity to be ingested by another host and repeat its life cycle. Some pathogens are shed in great numbers, with the random chance that a few organisms will find a new host to repeat the life cycle. In either case, the survival of the pathogen outside its host is dependent on many variables, including temperature, soil type, and moisture content. Table 6.7 shows the survival of different pathogens in different environment. Pathogens can be inactivated after exposure to the environment, but may survive long enough to be of public and/or animal health concern (Venglovsky et al. 2006). In pig slurry, *S. senftenberg* is reported to survive for 14 d at 8 °C, 8 d at 20 °C, and 8 d at 37 °C (Guan and Holley 2003). In cattle slurry, *Salmonella enterica* Typhimurium was not detectable after 19 d at 37 °C, but survived for at least 60 d at 4 and 20 °C (Himathongkham et al. 1999). Guan and Holley (2003) reported that *Salmonella enterica* Typhimurium and *E. coli* O157:H7 survived less well in cattle liquid manure than in solid manure at 20 and 37 °C while the opposite was true at 4 °C. Therefore, handling manure as a liquid may be a better alternative as less time is required to kill the pathogens.

Table 6.7: Survival times (days*) of pathogens in different media (modified from Guan and Holley, 2003)

Media	Temperature	*Giardia*	*Cryptosporidium*	*Salmonella*	*Campylobacter*	*E. coli O157:H7*	*Yersinia enterocolitica*
Water							
	Frozen (–4 °C)	< 7	> 84	> 182	14–56	> 300	> 365
	Cold (4–8 °C)	77	> 84	> 182	8–120	> 91	> 365
	Warm (20–30 °C)	14	70	45–152	< 2	49–84	10
Soil							
	Frozen (–4 °C)	< 7	> 84	> 84	14-56	> 300	> 365
	Cold (4–8 °C)	49	56	63	20	99	> 365
	Warm (20–30 °C)	7	28	> 45	10	56	10
Cattle Manure							
	Frozen (–4 °C)	< 7	> 84	> 196	14-56	> 100	> 365
	Cold (4–8 °C)	7	56	84–196	7-21	> 70	30–100
	Warm (20–30 °C)	7	28	48	3	49–56	10–30
Liquid Pig Manure		365	> 365	13–75	> 112	10–100	12–28
Compost		14	28	7–14	7	7	7
Dry Surface		1	1	1–7	1	1	1

* Days required until the pathogen can no longer be detected.

Solid manures are incorporated in the topsoil or applied as liquid or slurry. Both these methods have different implications in terms of pathogen survival in soil. For example, Niewolak (1994) stated that microorganisms (*Escherichia*, *Salmonella*) that reach the soil by means of application of contaminated pig slurry can penetrate to the depth of 160–180 cm. Deeper penetration of some pathogens may help their survival as the surface soil is more prone to environmental variations than the deeper soil. Henry et al. (1995) isolated *Salmonella* from pasture 2 m, and from soil 8 m, after the application of contaminated pig slurry. Similarly, soils amended with solid manures can infect the leafy (ready to eat) vegetables/sprout or fruits falling on the manure amended soils. The factors that may influence the pathogen survival in soil include temperature, soil type, soil moisture content, presence of aerobic microorganisms, soil levels of resin-extractable phosphorus, exchangeable aluminum, and soil pH. *E. coli* O157:H7 can survive in soil for a long period of time depending on the soil type. In the laboratory, the organism survived for at least 8 wk in moist soil at 25 °C (Mubiru et al. 2000). Under fluctuating environmental temperatures (–6.5 to 19.6 °C), the organism can be detected up to 99 d (Bolton et al. 1999). Among the physical and chemical properties of soil, soil moisture is a major factor determining bacterial survival. Greater survival is often associated with moist soils, thus rainfall is a factor that favors bacterial survival. Contrary to *E. coli*, field studies in which *Salmonella* contaminated manure or manure slurry was applied to soil have shown considerable variation in the length of time the pathogens survived (Zibilske and Weaver 1978). Survival of *Salmonella* ranged from a few days in one study to 2 years in another. The only set of conditions that consistently led to death of all the *Salmonellae* within 1 week was the incubation temperature of 39 °C and the dry soil condition. Zibilske and Weaver (1978) reported the survival of *Salmonella enterica* Typhimurium in soil for 42 d at 22 °C, and for 63 d at 5 °C. Under natural environmental conditions, *Salmonella enterica* Typhimurium was isolated up to 14 d from agricultural soil amended with *Salmonella* contaminated hog manure slurry, which was spread in spring (Baloda et al. 2001). Under controlled conditions in terrestrial ecosystems, *Salmonella enterica* Typhimurium, DT 104 and DT 12, has been reported to survive up to 299 d. Variation in the length of survival period in some studies was attributed to variation in sunlight intensity on test plots (Tannock and Smith 1971; Tannock and Smith 1972). In case of *Y. enterocolitica* bacteria, it survived for 7 d in soil at 30 °C under the laboratory conditions (Chao et al. 1988). Survival times could be prolonged to 10 d in pH-adjusted soil (final pH 6.5), and air drying of the pH-adjusted soil reduced the number of survivors. Similarly, the protozoa *Giardia* is sensitive to freezing of soil; *Cryptosporidium* is resistant. Olson et al. (1999) reported that *Giardia* cysts in soil were non-infective after 7 d at –4 °C, but *Cryptosporidium* could survive for >12 wk. Infective *Giardia* cysts and *Cryptosporidium* oocysts were recovered from soils maintained at 4 °C for up to 8 wk. At 25 °C, *Giardia* cysts were inactivated in 1 wk in soil, whereas *Cryptosporidium* oocysts survived and were infective for 4 wk. In cold (4 to 6 °C) soil, most pathogens can survive for at least 30 d. Helminths are perhaps the most persistent of enteric pathogens. *Ascaris* eggs survive several years in soils, although very dry or very wet soils decrease the survival (Straub et al. 1993; Varadyova et al. 2001). Intensive application of manures to soil has also been reported to affect the pathogen community in soil. Higher manure-to-soil ratio (e.g., 1 part manure to 10 parts soil versus 1:25, 1:50 or 1:100) has been reported to enhance the inactivation of *E. coli* O157:H7 in the soil (Jiang et al. 2002). Among all the pathogens considered, the greatest survival was found for *Cryptosporidium* in frozen soil, and *E. coli* O157:H7 and *Salmonella* in warm (20 to 30 °C) soil (Table 7.7).

The soil native microorganisms also influence the survival of any alien pathogen community. Microorganisms naturally occurring in the soil appear to have a major influence on the rates of inactivation of *E. coli* O157:H7 at 15 and 21 °C, as the rates of pathogen inactivation were considerably less at these temperatures in manure amended autoclaved soil than in manure-amended unautoclaved soil (Jiang et al. 2002). In an early study, *Campylobacter intestinalis* survived in nonsterilized soil for 20 and 10 d at 6 °C and 20 to 37 °C, respectively. It has been suggested that soil sterilization destroys competing indigenous microflora, increases available nutrients for the microorganisms, and reduces the concentrations of inhibitory compounds unstable at treatment temperatures (Guan and Holley 2003).

Contamination of Freshwater Resources with Pathogens

Agriculture is often attributed as the source of water quality impairments, with both nutrients and pathogens as pollutants. Land applications of raw manure result in the contamination of agricultural runoff and water supplies. Contaminated drinking water has the potential to cause extensive outbreaks due to the large populations served by many distribution systems. Water contaminated by the pathogens is often a health risk for swimmers and is potentially hazardous for other recreational uses. It must receive extensive treatment. Studies have shown that the oocysts of *Cryptosporidium* spp. were present in 39 to 87 percent of surface water (rivers, lakes) tested throughout the United States from 1988 to 1993 (LeChevallier et al. 1991; LeChevallier and Norton 1995). The sources of the organism are typically unknown (Juranek 1995; Atwill 1996; Sterling 1990; Smith and Rose 1990). Results showed that significant numbers of bacteria could reach the surface water by infiltrating through the soil and traveling through subsurface tile drains to the receiving water (Joy et al. 1998). Similarly, Goss et al. (1998) examined well contamination problems resulting from the use of manure in Ontario, and reported that the number of wells showing bacterial contamination increased from 15 to 25% between 1955 and 1992.

Water plays a major role in mobilizing and transporting microorganisms. Rainfall washes organisms from feces or vegetation surfaces and directs them into the soil or along the land surface and thus into surface water. Deliberate discharge, spillage (e.g. from storage, during transport, or during processing and usage), leakage and runoff (e.g. of agricultural chemicals) are all important in terms of aqueous pollutants. Legal limits for discharge to streams are set for many industries, aimed at keeping the level of contamination within the accepted limits. Illegal discharges, or accidental spillage, however, sometimes occur and accounted for the majority of reported surface water pollution incidents in some countries (Briggs 2003). Dumping (both legally in landfill sites and illegally) represents a major source of emission of solid wastes, though final release into the wider environment may only occur when these materials decompose or break up. Landfill sites are thus responsible for the emission of a wide range of pollutants, via different pathways, especially when these sites are inadequately sealed or poorly maintained (Rushton 2003).

Once a pathogen leaves the host environment, it must adjust to external conditions that are different and often stressful. Biological, chemical, and physical stressors interact in complex ways that have not been well studied. The survival of most pathogens, once discharged into a water body, is highly variable depending upon the quality of the receiving water, particularly turbidity, oxygen levels, presence of pesticides and nutrients, temperature, pH, organic matter content, and solar radiation (Moore et al. 1988). Once in a water body, microorganisms often get adsorbed to the organic matter and soil particles. These settle out and accumulate at the bottom of rivers and lakes, and may become a source of organism if re-suspended. Many small protozoa feed on bacteria, including pathogens, and many invertebrates feed on both bacteria and protozoa in a water body.

The survival of several major zoonotic pathogens under different environmental conditions is summarized in Table 2.7. In general, zoonotic pathogens appear to survive longer in the water, followed by soil and manure. In each of these environments, they survive better at lower than at higher temperatures. Low temperatures, which slow the metabolism, generally prolong the survival of pathogens. Most bacterial pathogens are sensitive to temperatures exceeding 60 °C (Rosen 2000). Bacterial pathogens, which produce resistant endospores or have thick-walled cells, may survive higher temperatures and are only killed by prolonged heating at temperatures greater than 100 °C (212 °F) (Kelly 1978). Higher temperatures also kill protozoan cysts. Similarly, pathogen mortality also results when the organisms are subjected to repeated freezing and thawing cycles.

Extensive research on coliforms, such as *E. coli*, provides the basis for understanding the factors affecting bacterial survival in streams, rivers, and lakes. Wang and Doyle (1998) reported that the survival of *E. coli* O157:H7 was higher in filtered, autoclaved municipal water and least in lake water. Regardless of the water source, survival was greatest at 8 °C and least at 25 °C. The pathogen survived for at least 91 d at 8 °C, but was not detectable within 49 to 84 d at 25 °C. It was also demonstrated that *E. coli* O157:H7 can enter a viable but nonculturable (VBNC) state in water (Wang and Doyle 1998). This VBNC state of the pathogen in the natural environment poses an epidemiological concern, since the source of contamination can be overlooked during outbreak investigations where conventional recovery methods are used. The fate of *Salmonella* species (including *Salmonella enterica* Typhimurium DT 104) in river water was examined by Santo Domingo et al. (2000). The authors inoculated river water and steam sterilized (autoclaved) river water with 8 $\log_{10}$ colony forming units (CFU) ml^{-1}

Salmonella. After 45 d at room temperature, they observed that the number of *Salmonella* was reduced by more than 3 logs in the untreated water, while the number of these organisms was reduced by 2 to 3 logs in the treated water.

Lengthy survival of *Yersinia enterocolitica* in water was noted by Karapinar and Gonul (1991), who detected survivors in sterile spring water after 64 wk at 4 °C. Survival was greatly reduced with increasing temperatures. In water, *Campylobacter* survived for 8 to 28 d at 4 °C (Buswell et al. 1998; Terzieva and McFeters 1991). Survival times were similar at 16 °C (Terzieva and McFeters 1991), but were greatly shortened as the temperature increased to 22 °C and above. At 22 and 37 °C, the organism only survived for 43 and 22 h, respectively (Buswell et al. 1998). *Giardia* cysts are much more susceptible to environmental stress than *Cryptosporidium* oocysts. A study by Olson et al. (1999) showed that temperatures as low as –4 °C inactivated *Giardia* cysts in water while *Cryptosporidium* oocysts remained viable for >12 wk. At 4 °C in water, *Giardia* cysts were infective for 11 wk, while *Cryptosporidium* oocysts again survived for >12 wk. In cold water (4 to 8 °C), *Y. enterocolitica* has the greatest survival among all the pathogens, whereas *Campylobacter* has possibly the least. In warm water (20 to 30 °C), *Salmonella* survives best, followed by *E. coli* O157:H7, *Cryptosporidium*, *Giardia*, *Y. enterocolitica*, and lastly *Campylobacter*.

Apart from the temperature, nutrient level and native microorganisms also affect the survival of pathogens in water. Korhonen and Martikainen (1991) reported that native bacterial flora outcompete the pathogens for nutrients. In laboratory studies of lake water, *E. coli* is reported to survive for prolonged periods (as long as 260 days) when kept at 4 °C (Flint, 1987) when no other bacteria was present. Nutrient enrichment of water may play an important role in the survival. Lim and Flint (1989) demonstrated the importance of nitrogen in the survival of *E. coli*. They allowed the cells to survive the competition from the indigenous bacterial flora. The additional nitrogen appears to be important in allowing cells to go through a period of progressive cell dormancy that prolongs their viability.

REFERENCES

Adak, G.K., Long, S.M. and O'Brien, S.J. (2002). Trends in indigenous food-borne diseases and deaths, England and Wales: 1992 to 2000. *Gut*. **51:** 832-841.

Alexandratos, N. (1998). World food and agriculture outlook to year 2010. Proceeding of the FAO/IFA/IFDC informal meeting on fertilizer use by crop. 2-3 March, Rome, Italy.

Atwill, E.R. (1996). Assessing the link between rangeland cattle and waterborne *Cryptosporidium parvum* infection in humans. *Rangelands* **18:** 48-51.

Ayoub, A.T. (1999). Fertilizers and the environment. *Nutrient Cycling in Agroecosystems* **55:** 117-121.

Baloda, S.B., Christensen, L. and Trajcevska, L. (2001). Persistence of a *Salmonella enterica serovar* Typhimurium DT12 clone in a piggery and agricultural soil amended with *Salmonella*-contaminated slurry. *Appl. Environ. Microbiol.* **67:** 2859-2862.

Barker, T., Bashmakov, I., Bernstein, L., Bogner, J.E. and Bosch, P.R. et al. (2007). Technical Summary. *In:* Metz, B., Davidson, O.R., Bosch, P.R., Dave, R. and Meyer, L.A. (Eds.).

Climate Change 2007: Mitigation. Contribution of Working Group III to the Fourth Assessment Report of the Intergovernmental Panel on Climate Change Cambridge University Press, Cambridge, United Kingdom and New York, NY, USA.

Behrendt, H. (1996). Inventories of point and diffuse sources and estimated nutrient loads — a comparison for different river basins in central Europe. *Water Sci. Technol.* **33:** 99-107.

Besser, R.E., Lett, S.M., Weber, J.T., Doyle, M.P., Barett, T.J., Wells, J.G. and Griffin, P.M. (1993). An outbreak of diarrhea and hemolytic uremic syndrome resulting from *E. coli* O157:H7 in fresh pressed apple cider. *JAMA J. Am. Med. Assoc.* **269:** 2217-2220.

Bitton, G. (1994). Wastewater Microbiology. Wiley-Liss, Inc., New York.

Bockman, O.C., Kaarstad, O., Lie, O.H. and Richards, I. (1990). Agriculture and Fertilizers. Agricultural Group, Norsk Hydro a.s, Oslo, Norway.

Bolton, D.J., Byrne, C.M., Sheridan, J.J., McDowell, D.A., Blair, I.S. and Hegarty, T. (1999). The survival characteristics of a non-pathogenic strain of *E. coli* 0157:H7. *In:* Duffy, G., Garvey, P., Coia, J., Wasteson, Y. and McDowell, D.A. (Eds.). Verocytotoxigenic *E. coli* in Europe. 2. Survival and growth of verocytotoxigenic *E. coli.* Teagasc. The National Food Centre, Dublin, Ireland. pp. 28-36.

Bradley, J.W., Shunsoku, K., Matthews, P., Sato, K. and Webber, M. (1992). Worldwide sludge management practices. *In:* Leu-Hing, C., Zenz, D.R. and Kuchenrither, R. (Eds.). Water Quality Management Library: Municipal Sewage Sludge Management: Processing, Utilization and Disposal, Vol. 4. Technomic Publishing Company, Lancaster, Pennsylvania, pp. 537-658.

Breuer, T., Benkel, D.H., Shapiro, R.L., Hall, W.N., Winnett, M.M., Linn, M.J., Neimann, J., Barrett, T.J., Dietrich, S., Downes, F.P., Toney, D.M., Pearson, J.L., Rolka, H., Slutsker, L. and Griffin, P.M. (2001). A multistate outbreak of *E. coli* O157:H7 infections linked to alfalfa sprouts grown from contaminated seeds. *Emerg. Infect. Dis.* **7:** 977-982.

Briggs, D. (2003). Environmental pollution and the global burden of disease. *British Medical Bulletin* **68:** 1-24.

Bumb, B.L. and Baanante, C.A. (1996). The role of fertilizers in food security and the environment: Trends to 2020. Discussion paper No. 17, International Food Policy Research Institute (IFPRI). Washington, DC, USA.

Bundy, L.G. (2001). Managing urea-containing fertilizers. University of Wisconsin. Presentation at 2001 Area Fertilizer Dealer Meetings, 27 Nov-6 Dec. 2001. Available at http://www.soils.wisc.edu/extension/publications/horizons/2001/Urea_management.pdf [Accessed 3 Nov. 2008].

Buswell, C.M., Herlihy, Y.M., Lawrence, L.M., McGuiggan, J.T.M., Marsh, P.D., Keevil, C.W. and Leach, S.A. (1998). Extended survival and persistence of *Campylobacter* spp. in water and aquatic biofilms and their detection by immunofluorescent-antibody and -rRNA staining. *Appl. Environ. Microbiol.* **64:** 733-741.

Carpenter, S.R., Caraco, N.F., Correll, D.L., Howarth, R.W., Sharpley, A.N. and Smith, V.H. (1998). Nonpoint pollution of surface waters with phosphorus and nitrogen. *Ecol. Appl.* **8(3):** 559-568.

Chao, W.L., Ding, R.J. and Chen, R.S. (1988). Survival of *Yersinia enterocolitica* in the environment. *Can. J. Microbiol.* **34:** 753-756.

Chapman, P.J. and Edwards, A.C. (1999). The impact of atmospheric nitrogen on the behaviour of nitrogen in surface waters. *In:* Langan, S.J. (Ed.). The impact of nitrogen deposition on natural and semi-natural ecosystems, Kluwer Academic Publishers, Dordrecht, The Netherlands. pp. 153-212.

Department for Environment, Food and Rural Affairs (2002). Sewage sludge: Proposals to amend the statutory controls for agricultural use (consultation paper) DEFRA, London.

Doyle, M.P. (2000). Reducing food-borne disease: what are the priorities? *Nutrition* **16:** 647-649.

Dyson, T. (1998). World food trends and prospects to 2015. NAS Colloquium on plants and population: Is there time? Available at http://146.186.172.254/nas/Speakers/Dyson%20 manuscript.html [Accessed 9 Dec. 2008].

Edwards, A.C. and Withers, P.J.A. (1998). Soil phosphorus management and water quality: A UK perspective. *Soil Use Manage* **14:** 124-130.

Emmett, W.W. (1980). Overland flow. *In:* Kirkby, M.H. (Ed.). Hillslope Hydrology. John Wiley and Sons. pp. 145-176.

FAO (2000). Fertilizer requirements in 2015 and 2030. Available at www.fao.org [Accessed 2 Feb. 2008].

FAO (2003). World Agriculture: Towards 2015/2030. An FAO Perspective. FAO, Rome.

FAO (2008a). FAOSTAT Available at http://faostat.fao.org/ [Accessed 24 January 2008].

FAO (2008b). Current world fertilizer trends and outlook to 2012. Available at ftp://ftp.fao.org/agl/agll/docs/cwfto11.pdf [Accessed 20 Dec. 2008].

FAOSTAT (2006). FAOSTAT Agricultural Data. Available at http://faostat.fao.org [Accessed 16 Dec. 2008].

Feachem, R.G., Bradley, D.J., Garelick, D.J. and Mara, D.D. (1983). Sanitation and Disease: Health Aspects of Excreta and Wastewater Management. John Wiley and Sons, New York.

Fennesy, M.S. and Cronk, J.K. (1997). The effectiveness and restoration potential of riparian ecotones for the management of nonpoint source pollution, particularly nitrate. *Crit. Rev. Environ. Sci. Technol.* **27:** 285-317.

Filip, Z., Kaddu-Mulindwa, D. and Milde, G. (1988). Survival of some pathogenic and facultative pathogenic bacteria in groundwater. *Water Sci. Technol.* **20:** 227-231.

Flint, K.P. (1987). The long-term survival of *E. coli* in river water. *J. Appl. Bacter.* **63:** 261-270.

Freeman, T.M. and Cawthon, D.L. (1999). Use of composted dairy cattle solid biomass, poultry litter and municipal biosolids as greenhouse growth media. *Compost Sci. Utilization* **3:** 66-71.

Friedman, M.S., Roels, T., Koehler, J.E., Feldman, L., Bibb, W.F. and Blake, P. (1999). *E. coli* O157:H7 Outbreak associated with an improperly chlorinated swimming pool. *Clin. Infect Dis.* **29(2):** 298-303.

Frink, C.R., Waggoner, P.E. and Ausubel, J.H. (1999). Nitrogen fertilizer: Retrospect and prospect. *Proc. Natl. Acad. Sci.* **96:** 1175-1180.

Gaines, T.P. and Gaines, S.T. (1994). Soil texture effect on nitrate leaching in soil percolates. Commun. *Soil Sci. Plant Anal.* **25(13&14):** 2561-2570.

Galloway, J.N. and Cowling, E.B. (2002). Reactive nitrogen and the world: 200 years of change. *Ambio* **31:** 64-71.

Galloway, J.N., Aber, J.D., Erisman, J.W., Seitzinger, S.P., Howarth, R.H., Cowling, E.B. and Cosby, B.J. (2003). The nitrogen cascade. *BioScience* **53:** 341-356.

Galloway, J.N. (2003). The global nitrogen cycle. *Treatise on Geochemistry,* **8:** 557-583.

Gifford, G. and Busby, F. (1973). Loss of particulate organic materials from semi-arid watersheds as a result of extreme hydrologic events. *Water Resour. Res.* **9:** 1443-1449.

Goss, M.J., Barry, D.A.J. and Rudolph, D.L. (1998). Contamination in Ontario farmstead domestic wells and its association with agriculture: Results from drinking water wells. *J. Contam. Hydrol.* **32:** 267-293.

Griffiths, M.W. (2000). The new face of food-borne illness. CMSA News. 6-9 March. Canadian Meat Science Assoc., Ottawa, ON.

Grigg, D.B. (1974). Agricultural Systems of the World. Cambridge University Press. pp. 45-55.

Guan, T.Y. and Holley, R.A. (2003). Pathogen survival in swine manure environments and transmission of human enteric illness—A review. *J. Environ. Qual.* **32:** 383-392.

Henry, D.P., Frost, A.J., O'Boyle, D. and Cameron, R.D.A. (1995). The isolation of salmonellas from piggery waste water after orthodox pondage treatment. *Aust. Vet. J.* **72:** 478-479.

Himathongkham, S., Bahari, S., Riemann, H. and Cliver, D. (1999). Survival of *Escherichia coli* O157:H7 and *Salmonella typhimurium* in cow manure and cow manure slurry. *FEMS Microbiol. Lett.* **178:** 251-257.

Howarth, R.W., Anderson, D., Cloern, J., Elfring, C., Hopkinson, C. and Lapointe, B. (2000). Nutrient pollution of coastal rivers, bays, and seas. *Iss. Ecol.* **7:** 1-15.

IMF (2008). World Economic Outlook, Financial Stress, Downturns, and Recoveries. Available at http://www.imf.org/external/pubs/ft/weo/2008/02/ [Accessed 17 Dec. 2008].

Jiang, X., Morgan, J. and Doyle, M.P. (2002). Fate of *Escherichia coli* O157:H7 in manure-amended soil. *Appl. Environ. Microbiol.* **68(5):** 2605-2609.

Joy, D.M., Lee, H., Reaume, C.M., Whiteley, H.R. and Zelin, S. (1998). Microbial contamination of subsurface tile drainage water from field applications of liquid manure. *Can. Agric. Eng.* **40:** 153-160.

Juranek, D.D. (1995). Cryptosporidiosis: sources of infection and guidelines for prevention. *Clin. Inf. Dis.* **21(Suppl. 1):** S57-S61.

Karapinar, M. and Gonul, S.A. (1991). Survival of *Yersinia enterocolitica* and *Escherichia coli* in spring water. *Int. J. Food Microbiol.* **13:** 315-320.

Karl-Thomas, R. and Trenberth, K.E. (2003). Modern Global Climate Change. *Science* **302:** 1719-1723.

Kelly, W.R. (1978). Animal and human health hazards associated with the utilization of animal effluents. *In:* A Workshop in the EEC Programme of Coordination of Research on Effluents, November 21-23, Dublin, Ireland. Commission of the European Communities, Luxembourg.

Korhonen, L. and Martikainene, P.J. (1991). Survival of *E. coli* and *Campylobacter jejuni* in untreated and filtered lake water. *J. Appl. Bacter.* **71:** 379-382.

Kotak, B.G., Kenefick, S.L., Fritz, D.L., Rousseaux, C.G., Prepas, E.E. and Hrudey, S.E. (1993). Occurrence and toxicological evaluation of cyanobacterial toxins in Alberta lakes and farm dugouts. *Water Res.* **27:** 495-506.

Kroeze, C. and Seitzinger, S.P. (1998). Nitrogen inputs to rivers, estuaries, and continental shelves and related nitrous oxide emissions in 1990 and 2050: a global model. *Nut. Cy. Agroecosyst.* **52:** 195-212.

Kuhlmann, H. and Engels. T. (1989). Nitrogen utilisation in relation to N-fertilisation. The Fertiliser Society, London.

Laegrid, M., Bockman, O.C. and Kaarstad, O. (1999). Concerns Related to Fertilizer Use. *In:* Agriculture, Fertilizers and the Environment. CABI Publishing. pp. 79-182.

Lawton, L.A. and Codd, G.A. (1991). Cyanobacterial (blue-green algae) toxins and their significance in UK and European waters. *J. Inst. Water Environ. Manage* **5:** 460-465.

Le Treut, H., Somerville, R., Cubasch, U., Ding, Y., Mauritzen, C., Mokssit, A., Peterson, T. and Prather, M. (2007). Historical Overview of Climate Change Science *In:* Solomon, S., Qin, D., Manning, M., Chen, Z., Marquis, M., Averyt, K.B., Tignor, M. and Miller, H.L. (Eds.). Climate Change 2007: The Physical Science Basis. Contribution of Working Group I to the Fourth Assessment Report of the Intergovernmental Panel on Climate Change. Cambridge University Press, Cambridge, United Kingdom and New York, NY, USA.

LeChevallier, M.W. and Norton, W.D. (1995). *Giardia* and *Cryptosporidium* in raw and finished water. *J. Am. Water Works Assoc.* **87:** 54-68.

LeChevallier, M.W., Norton, W.D. and Lee, R.G. (1991). Occurrence of *Giardia* and *Cryptosporidium* sp. in surface water supplies. *Appl. Environ. Microbiol.* **57:** 2610-2616.

Liang, T., Zhang, X., Zhang, S., Xingxiu, Y.U. and Wang, H. (2002). Nitrogen export by runoff and sediment under different types of land use in West Tiaoxi catchment. *J. Geogr. Sci.* **12(4):** 443-450.

Lim, C.H. and Flint, K.P. (1989). The effects of nutrients on the survival of *E. coli* in lake water. *J. Appl. Bacter.* **66:** 559-569.

Lowrance, R. and Williams, R. (1988). Carbon movement in runoff and erosion under simulated rainfall conditions. *Soil Sci. Soc. Am. J.* **52:** 1445-1448.

Marxsen, C.S. (2001). Potential world garbage and waste carbon sequestration. *Environ. Sci. Policy* **4:** 293-300.

Mead, P. and Griffin, P.M. (1998). *Escherichia coli* O157:H7. *Lancet* **352:** 1207-1212.

Mead, P.S., Slutsker, L., Dietz, V., McCaig, L.F., Bresee, J.S., Shapiro, C., Griffin, P.M. and Tauxe, V.R. (1999). Food-related illness and death in the United States. *Emerg. Infect. Dis.* **5(5):** 607-625.

Merz, B. and Bárdossy, A. (1998). Effects of spatial variability on the rainfall runoff process in a small loess catchment. *J. Hydrol.* **212 & 213:** 304-317.

Miao, X.H. and Zhao, W.C. (2002). The situation of livestock pollution and the integrated utilization of animal manure in Gansu province. *Environ. Res. Inspector. Gansu Province* **15(4):** 305-306.

Miner, J.R., Humenik, F.J. and Overcash, M.R. (2000). Managing livestock wastes to preserve environmental quality. Iowa State University Press, Ames, Iowa.

Moore, J.A., Smyth, J., Baker, S. and Miner, J.R. (1988). Evaluating coliform concentrations in runoff from various animal waste management systems. Special Report 817. Agricultural Experimental Stations Oregon State Univ., Corvallis, and U.S. Dep. Agric., Portland, OR.

Moore, R.S., Taylor, D.H., Sturman, L.S. and Reddy, M.M. (1981). Poliovirus adsorption by 34 minerals and soils. *Appl. Environ. Microbiol.* **42:** 963-975.

Mosier, A.R. (2002). Environmental challenges associated with needed increase in global nitrogen fixation. *Nutr. Cycl. Agroecosy* **63:** 101-116.

Mosier, A.R., Duxbury, J.M., Freney, J.R., Heinemeyer, O., Minami, K. and Johnson, D.E. (1998). Mitigating agricultural emissions of methane. *Climatic Change* **40:** 39-80.

Mubiru, D.N., Coyne, M.S. and Grove, J.H. (2000). Mortality of *Escherichia coli* O157:H7 in two soils with different physical and chemical properties. *J. Environ. Qual.* **29:** 1821-1825.

Niewolak, S. (1994). The sanitary and bacteriological estimation of surface and underground waters in the region of pig slurry agricultural utilization. *Pol. J. Environ. Stud.* **3:** 31-37.

Nixon, S.W. (1995). Coastal marine eutrophication: definition, causes, and future concerns. *Ophelia,* **41:** 119-219.

Noah, G.F. and Emmanuel, J.G. (2002). Carbon and nitrogen losses by surface runoff following changes in vegetation. *J. Environ. Qual.* **31:** 1207-1213.

Norse, D. (2003). Fertilizers and world food demand—implications for environmental stresses. *Agro-Chemicals Report* **3(2):** 13-22.

NRC (1992). Restoration of aquatic ecosystems: science, technology and public policy, National Academy Press, Washington, DC, USA.

Olson, M.E., Goh, J., Philips, M., Guselle, N. and McAllister, T.A. (1999). *Giardia* cyst and *Cryptosporidium* oocyst survival in water, soil, and cattle feces. *J. Environ. Qual.* **28:** 1991-1996.

Oram, P. and Hojjati, B. (1995). The growth potential of existing agricultural technology. *In: Nurul Islam* (Ed.). Population and feed in the early Twenty-first century. International food policy research institute (IFPRI), Washington, DC.

Palmstrom, N.S., Carlson, R.E. and Cooke, G.D. (1988). Potential links between eutrophication and formation of carcinogens in drinking water. *Lake Reserv. Manage* **4:** 1-15.

Patrick, H. and Prud'homme, M. (2008). Summary Report of 34th IFA Enlarged Council Meeting; World Agriculture and Fertilizer Demand, Global Fertilizer Supply and Trade 2008-2009. A/08/158, Ho-Chi-Minh, Viet Nam.

Paustian, K., Babcock, B.A., Hatfield, J., Lal, R., McCarl, B.A., McLaughlin, S., Mosier, A., Rice, C., Robertson, G.P., Rosenberg, N.J., Rosenzweig, C., Schlesinger, W.H. and Zilberman, D. (2004). Agricultural Mitigation of Greenhouse Gases: Science and Policy Options. CAST (Council on Agricultural Science and Technology) Report, R141 2004, ISBN 1-887383-26-3.

Pell, A.N. (1997). Manure and Microbes: Public and Animal Health Problem? *J. Dairy Sci.* **80:** 2673-2681.

Plucknett, D. (1995). Prospects of meeting future food needs through new technology. *In:* Nurul Islam, (Ed.). Population and feed in the early Twenty-first century. International food policy research institute (IFPRI), Washington, DC.

Riley, L.W., Remis, R.S., Helgerson, S.D., McGee, H.B., Wells, J.G., Davis, B.R., Hebert, R.J., Olcott, E.S., Johnson, L.M., Hargrett, N.T., Blake, P.A. and Cohen, M.L. (1983). Hemorrhagic colitis associated with a rare *E. coli* serotype. *New England J. Medicine* **308:** 681-685.

Robert, H. (2004). Predicting and measuring nitrogen loss. University of IL Pest Mgmt Newsletter. 28 May 2004. Available at http://www.ipm.uiuc.edu/bulletin/article.php? issueNumber=10&issueYear=2004&articleNumber=8 [Accessed 19 Jan 2009].

Rogner, H.H., Zhou, D., Bradley, R., Crabbé, P., Edenhofer, O., Hare, B., Kuijpers, L. and Yamaguchi, M. (2007). Introduction. *In:* Metz, B., Davidson, O.R., Bosch, P.R., Dave, R. and Meyer, L.A. (Eds.). Climate Change 2007: Mitigation. Contribution of Working Group III to the Fourth Assessment Report of the Intergovernmental Panel on Climate Change. Cambridge University Press, Cambridge, United Kingdom and New York, NY.

Rosegrant, M., Paisner, M.S. and Meijer, S. (2001). Long-Term Prospects for Agriculture and the Resource Base. The World Bank Rural Development Family. Rural Development Strategy Background Paper No. 1. The World Bank, Washington.

Rosen, B.H. (2000). Waterborne Pathogens in Agricultural Watersheds. WSSI-Technical Note 2.

Roy, R.N., Misra, R.V. and Montanez, A. (2002). Decreasing reliance on mineral nitrogen - yet more food. *Ambio* **31:** 177-183.

Sampat, P. (2000). Deep Trouble: The hidden threat of groundwater pollution. *In:* Peterson, J. (Ed.). Worldwatch Paper 154. Available at www.worldwatch.org [Accessed 29 Dec. 2008].

Santo, D.J.W., Harmon, S. and Bennett, J. (2000). Survival of *Salmonella* species in river water. *Curr. Microbiol.* **40:** 409-417.

Schindler, D.W. (1977). Evolution of phosphorus limitation in lakes. *Science* **195:** 260-262.

Seitzinger, S.P. and Kroeze, C. (1998). Global distribution of nitrous oxide production and N inputs in freshwater and coastal marine ecosystems. *Global Biogechem. Cycle* **12:** 93-113.

Shaffer, M. (2001). The threat of toxic fertilizer. California Public Interest Research Group Charitable Trust. San Francisco, CA.

Sharpe, R.R., Schomberg, H.H., Harper, L.A., Endale, D.M., Jenkins, M.B. and Franzluebbers, A.J. (2004). Ammonia volatilization from surface-applied poultry litter under conservation tillage management practices. *J. Environ. Qual.* **33:** 1183-1188.

Sharpley, A.N. and Smith, S.J. (1994). Wheat tillage and water quality in the Southern Plains. *Soil Till Res.* **30:** 33-38.

Smil, V. (1999). Nitrogen in crop production: An account of global flows. *Glob. Biogeochem. Cyc.* **13:** 647-662.

Smith, H.V. and Rose, J.B. (1990). Water-borne cryptosporidiosis. *Parasitol. Today* **6:** 8-12.

Smith, J.E.J.R. and Perdek, M.J. (2003). Assessment and management of watershed microbial contaminants. *Crit. Rev. Environ. Sci. Technol.* **33:** 1-27.

Smith, K.A. and Conen, F. (2004). Impacts of land management on fluxes of trace greenhouse gases. *Soil Use Manage* **20:** 255-263.

Smith, P., Martino, D., Cai, Z., Gwary, D., Janzen, H. and Kumar, P. et al. (2007). Agriculture. *In:* Metz, B., Davidson, O.R., Bosch, P.R., Dave, R. and Meyer, L.A. (Eds.). Climate Change 2007: Mitigation. Contribution of Working Group III to the Fourth Assessment Report of the Intergovernmental Panel on Climate Change. Cambridge University Press, Cambridge, United Kingdom and New York, NY.

Smith, V.H. (1998). Cultural eutrophication of inland, estuarine, and costal waters. *In:* Pace, M.L. and Groffman, P.M. (Eds.). Successes, limitation and frontiers in ecosystem science. Springer-Verlag, New York, NY.

Steinfeld, H., Gerber, P., Wassenaar, T., Castel, V., Rosales, M. and De Haan, C. (2006). Livestock's 2006 Long Shadow: Environmental Issues and Options. Food and Agriculture Organization of the United Nations, Rome.

Sterling, C.R. (1990). Waterborne cryptosporidiosis. *In:* Dubey, J.P., Speer, C.A. and Fayer, R. (Eds.). Cryptosporidiosis of man and animals. CRC Press, Boca Raton, FL. pp. 51-58.

Straub, T.M., Pepper, I.L. and Gerba, C.P. (1993). Hazards from pathogenic microorganisms in land-disposed sewage sludge. *Rev. Environ. Contam. Toxicol.* **132:** 55-91.

Tannock, G.W. and Smith, J.M.B. (1971). Studies on the survival of *Salmonella typhimurium* and *Salmonella bovismorbificans* on pasture and in water. *Aust. Vet. J.* **47:** 557-559.

Tannock, G.W. and Smith, J.M.B. (1972). Studies on the survival of *Salmonella typhimurium* and *Salmonella bovismorbificans* on soil and sheep feces. *Res. Vet. Sci.* **13:** 150-153.

Taormina, P.J., Beuchat, L.R. and Slutsker, L. (1999). Infections associated with eating seed sprouts: An international concern. *Emerg. Infect Dis.* **5:** 626-634.

U.S. Department of Agriculture, (2004). Poultry and Eggs: Background. Available at http://www.ers.usda.gov/Briefing/Poultry/background.htm [Accessed 1 Dec. 2005].

UNEP/UNIDO/IFA (1996). Mineral Fertilizer Production and the Environment. A Guide to Reducing the Environmental Impact of Fertilizer Production. UNEP Technical Report No. 26.

UNIDO, IFDC (1998). Fertilizer Manual. 3rd edition. Kluwer Academic Publishers.

USDA (1992). National engineering handbook: Agricultural waste management field handbook. Part 651 (210-AWMFH, 4/92). Chapters 4 and 11. USDA, Washington, DC.

USEPA (2006). Global Anthropogenic Non-CO_2 Greenhouse Gas Emissions: 1990-2020. United States Environmental Protection Agency, EPA 430-R-06-003, June 2006. Washington, DC.

USEPA (1998). Environmental impacts of animal feeding operations. USEPA Office of Water, Standards and Applied Sci. Div., Washington, DC.

Varadyova, Z., Zelenak, I., Siroka, P. and Dubinsky, P. (2001). *In vitro* fermentation of cellulosis amorphous and meadow hay in experimentally *Ascaris suum* infected lambs. *Small Rumin Res.* **40(2):** 155-164.

Venglovsky, J., Martinez, J. and Placha, I. (2006). Hygienic and ecological risks connected with utilization of animal manures and biosolids in agriculture. *Livest. Sci.* **102:** 197-203.

Vitousek, P.M., Aber, J.D., Howarth, R.W., Likens, G.E., Matson, P.A., Schindler, D.W., Schlesinger, W.H. and Tilman, D.G. (1997). Human Alteration of the Global Nitrogen Cycle: Causes and Consequences. *Ecol.* **7(3):** 737-750.

Wang, G. and Doyle, M.P. (1998). Survival of enterohemorrhagic *E. coli* O157:H7 in water. *J. Food Prot.* **61:** 662-667.

Weber, C.A. (1990). Aufbau and vegetatioin der moore norddeutschlands. Bot Jahrb 40. *Beibl.* **90:** 19-34.

Whitehead, D.C. (1995). Grassland Nitrogen, CAB international, Wallingford.

Wilson, D. (1997). Fear in the Fields. The Seattle Times, July 3, Citing Agency for Toxic Substances Disease Registry, EPA.

WRI/UNEP/UNDP/WB (1996). World Resources: A Guide to the Global Environment, 1996-97, Oxford University Press, UK.

Zahoor, A., Honna, T. and Sadahiro Yamamoto, (2008). Leachability and phytoavailability of NPK from different bio-composts under chloride and sulphate dominated irrigation water. *J. Environ. Qual.* **37(3):** 1288-1298.

Zahoor, A., Honna, T., Yamamoto, S., Irshad, M., Haytham, E.S. and Abou El-Hassan, W.H., Faridullah, (2007). Wheat (*Triticum aestivum* L.) response to combined organic and inorganic phosphorus fertilizers application under saline conditions. *Acta. Agri. Scand B Plant Soil Sci.* **57(3):** 222-230.

Zibilske, L.M. and Weaver, R.W. (1978). Effect of environmental factors on the survival of *Salmonella typhimurium* in soil. *J. Environ. Qual.* **7(4):** 593-597.

7

Nitrate Management Approaches in Leafy Vegetables

SHAHID UMAR[1], ANJANA[2], NASER A. ANJUM[3] AND NAFEES A. KHAN[4]

[1]Department of Botany, Faculty of Science, Hamdard University, New Delhi – 110062, India.

[2]IGNOU Regional Centre, I.T.I. Building, Bulepur, Khanna, Distt. Ludhiana, Punjab – 141401, India

[3]CESAM-Centre for Environmental and Marine Studies & Department of Chemistry, University of Aveiro, 3810-193 Aveiro, Portugal

[4]Department of Botany, Faculty of Life Sciences, Aligarh Muslim University, Aligarh – 202002, U.P., India.
e-mail: s_umar9@hotmail.com

ABSTRACT

Achieving future food security for the burgeoning global population will require sustainable resource efficiency i.e., more must be grown on the same amount of land, using less water – with the help of a precise mix of nutrients. In terms of intensified use of varied fertilizers, excessive amounts of nitrogenous fertilizers are being applied to crops, considering that it is a reasonable insurance against yield losses and their economic consequences. Nevertheless, excess input of nitrogen more than its demand leads no or least nitrogen absorption by most of plants in addition to nitrogen buildup in the soil, mostly as nitrates. Moreover, as approximately 70% of total intake of dietary nitrate comes from vegetables and there has been a wide concern on the vegetable nitrate intake-mediated toxicity in human beings. A number of concerted efforts have been made to reduce the excessive N fertilizer use-mediated negative environmental impacts from agronomic to genetic engineering. To this end, the current chapter is aimed to overview the significance of plant nutrients for enhancing crop production, highlights nitrogen use in leafy vegetables production and critically discusses important strategies to be potentially useful for enhancing N-use efficiency in crops in general and vegetable crops in particular.

Keywords: Nitrate; Leafy vegetables; Toxicity; Health effects

INTRODUCTION

It is estimated that the world's population will reach >9 billion by 2050 that will require almost twice as much agricultural output – food, fiber and fuel – as today's 7 billion people (FAO 2010a,b; UN 2011). Feeding such burgeoning population has been expected to be relied mainly upon the availability of plant nutrients commensurate with the necessary increase in productivity, the deployment of new plant and farming technologies and the cultivation of more marginal land (Drescher et al. 2011). Additionally, achieving future food security will require resource efficiency i.e., more must be grown on the same amount of land, using less water – with the help of a precise mix of nutrients. It was estimated that in developing countries, 80 percent of the necessary production increases would come from increases in yields and cropping intensity and only 20 percent from expansion of arable land. Moreover, in terms of intensified use of varied fertilizers, excessive amounts of nitrogenous fertilizers are being applied to crops, considering that it is a reasonable insurance against yield losses and their economic consequences (Anjana et al. 2007a, 2007b). Nevertheless, excess input of nitrogen more than its demand leads no or least nitrogen absorption by most of plants in addition to nitrogen buildup in the soil, mostly as nitrates (Nosengo 2003). In fact, increased use of nitrogenous fertilizer has significantly caused imbalance of nutrients in the soil and increased the nitrate leaching to groundwater (NAAS 2005) and also altered the global nitrogen-cycle and produced nitrogenous gases of environmental consequence (Hakeem et al. 2012). According to Prakasa Rao and Puttanna (2000), nitrate contamination of groundwater is being intensified with the use of sewage water for irrigation of crops, which is prevalent throughout India. Moreover, as approximately 70% of total intake of dietary nitrate comes from vegetables and there has been a wide concern on the vegetable nitrate intake-mediated toxicity in human beings (Anjana et al. 2007b).

A number of concerted efforts including the split application of nitrogen, use of slow-release fertilizers, nitrification inhibitors, and the use of organic manures have been made to reduce the excessive nitrogen fertilizer use-mediated negative environmental impacts from agronomic to genetic engineering. However, the potential strategies, namely, the selection of nitrogen-efficient crops and/or adoption of strategies for enhancing nitrogen-use efficiency may largely work in isolation or combination in reducing ingestion of crop products with high nitrate-mediated anomalies in human. To this end, the current chapter overviews the significant plant nutrients for enhancing crop production, highlights nitrogen use in leafy vegetables production and critically discusses important strategies to be useful for enhancing nitrogen-use efficiency in crops in general and vegetable crops in particular.

LEAFY VEGETABLES AND NITRATES

Leafy vegetables are important elements of human diet. They are excellent sources of vitamins, minerals, sugars, folic acid and they do not have much calories. Some researchers have found that vegetable extract with chlorophyll is antimutagenic and anticarcinogenic and also has some antioxidant properties (Ma and Dolphin 1999). Vegetables are source of naturally occurring antioxidants, which plays an important role as a health protecting factors such as phenols, considered more powerful antioxidants than vitamin C or carotenoids to decrease disease risk (Larson 1988; Vinson et al. 1998). According to Caldwell and Britz (2006) chlorophyll and carotenoids have specific dietary activities and these pigments are sensitive to growing conditions. Moreover, everyday consumption of leafy vegetables lowers risk of cancer and heart diseases and prevents tiredness, (Ko³ton and Baran 2008).

To the brighter side, vegetables are the major sources of the daily intake of nitrate by human beings, supplying about 72 to 94% of the total intake (Dich et al. 1996). The leaves and roots of leafy green vegetables such as cauliflower, spinach, collard greens, lettuce, spinach and broccoli tend to contain the highest concentrations of nitrate, where it is absorbed from the soil and then subjected to the effects of photosynthesis. Nitrate concentration has been estimated by Anjana et al. (2007a) in different leafy vegetables, namely, spinach, fenugreek, chenopodium, coriander and sowa available in local markets of Delhi in which it was found that spinach and chenopodium contained fairly high concentration of nitrate that exceeded acceptable daily intake (ADI) limit for an average 60 kg person (if consumed @ 100 g day^{-1}) (Table 7.1). As mentioned also in Chapter 1 that when ingested within permissible limits, nitrate (NO_3^-) and nitrite (NO_2^-) ions play significant roles in various physiological functions of the human body including their roles in vascular and immune functions. However, excess nitrate and its metabolites are toxic and lead to various ailments in humans (Gupta et al. 1999, 2001; Sharma and Sharma 2012). Unfortunately, leafy vegetables grown under different agro-ecological conditions accumulate nitrate to potentially harmful concentrations. Hence, the greatest exposure to nitrate occurs from the consumption of leafy green vegetables (http://www.livestrong.com/article/446299-what-vegetables-have-nitrates/). Maximum NO_3^- levels in spinach leaves as proposed by the EFSA (2010) are 2500 mg kg^{-1} in fresh product for direct consumption, and 2000 mg kg^{-1} in fresh product for processing. In Switzerland, the maximum NO_3^- level allowed in vegetables is 3500 mg kg^{-1} in fresh product (Carranca et al. 2001). The European Union has prescribed the maximum limits for nitrate in lettuce and spinach (European Commission Regulation No. 1822/2005, Table 7.2). There are reports that a high nitrate accumulation in plants results in nitrite production, which is converted into nitric oxide (NO) which, together with O_2^-, could be rapidly catalyzed by nitrate reductase into peroxynitrite (ONOO–) which is highly toxic to plants (Durner and Klessig 1999; Lamattina et al. 2003; Anjana et al. 2007b). Therefore, high nitrate accumulation in plants is harmful to human health (Ikemoto et al. 2002; Ishiwata et al. 2002) as well as to plant growth (Reddy and Menary 1990; Anjana et al. 2007b). Therefore, in order to minimize the accumulation of nitrate in leafy vegetables and its ingestion by human beings, it is necessary to recognize major factors modulating the nitrate accumulation in leafy vegetables.

Table 7.1: Nitrate concentration in leafy vegetables collected from different markets in Delhi, India.

Market	Vegetable	No. of samples	Range (mg kg^{-1} fresh wt. of samples)†
Ghazipur	Spinach	15	**71**-1685
	Fenugreek	15	**288-524**
	Chenopodium	15	2794-**4451***
	Coriander	15	314-518
	Sowa	15	785-**1071**
	Petiole of spinach	15	462-887
Okhla	Spinach	15	1117-**4293***
	Fenugreek	15	208-331
	Chenopodium	15	**204**-510
	Coriander	15	**289**-486
	Sowa	15	**684**-847
	Petiole of spinach	15	5989-6269*
Azadpur	Spinach	15	2112-2892*
	Fenugreek	15	302-360
	Chenopodium	15	1420-2685*
	Coriander	15	388-**769**
	Sowa	15	866-934
	Petiole of spinach	15	2097-3568*

† Values are means of 15 samples
*Samples exceeding the ADI for an average 60 kg person (if consumed 100 g day^{-1})
Values in bold represent minimum and maximum concentrations in a given vegetable
Source: Anjana et al. (2007a)

Table 7.2: Summary of the maximum levels in European Commission (EC) Regulation No. 1822/2005.

Product	Harvest Period	Maximum level (mg NO_3/kg)
Fresh Spinach[1] (*Spinacia oleracea*)	1 October to 31 March	3000
	1 April to 30 September	2500
Preserved, deep-frozen or frozen spinach		2000
Fresh lettuce (*Lactuca sativa* L.) (protected and open-grown lettuce) excluding 'iceberg'-type lettuce	1 October to 31 March	
	-grown under cover	4500
	-grown in open air	4000
	1 April to 30 September	
	-grown under cover	3500
	-grown in open air	2500
'Iceberg'-type lettuces	Lettuce grown under cover	2500
Lettuce grown in open air		2000
Baby foods and processed cereal-based foods for infants and young children[2]		200

[1]The maximum levels for fresh spinach do not apply to fresh spinach to be subjected to processing and which is directly transported in bulk from the field to the processing plant.
[2]The maximum levels apply to the products as proposed ready for consumption or as reconstituted according to the instructions of the manufacturers.
Source: Anjana et al. (2007a)

Plant nitrate content is commonly viewed as an outcome of imbalance between its net absorption and assimilation rates (Cardenas-Navarro et al. 1999). The effect of endogenous and exogenous factors seems to be straightforward: on one hand, both uptake and assimilation systems are genetically determined (Ferrario-Mery et al. 1997; Ourry et al. 1997), explaining the variability of plant nitrate content among species and cultivars, whereas on the other hand, nitrate absorption depends on the availability of nutrition, and nitrate assimilation depends independently on climate, as it is partly a photosynthetic process in many plants (Ferrario-Mery et al. 1997). Alternatively, plant nitrate content might be either fixed through osmotic potential regulation (McIntyre 1997) or regulated per se through negative feedback on its transport systems (Cardenas-Novarro et al. 1999). In this perspective, Anjana et al. (2007b) suggested varied factors including the nutritional, environmental and physiological factors that significantly affect nitrate accumulation in plants. Among the various nutritional factors, potassium has been found to enhance the activities of nitrogen assimilating enzymes by stimulating the nitrate uptake by spinach plants, thereby enhancing nitrogen use efficiency of the plants (Anjana et al. 2009, Fig. 7.1; Table 7.3).

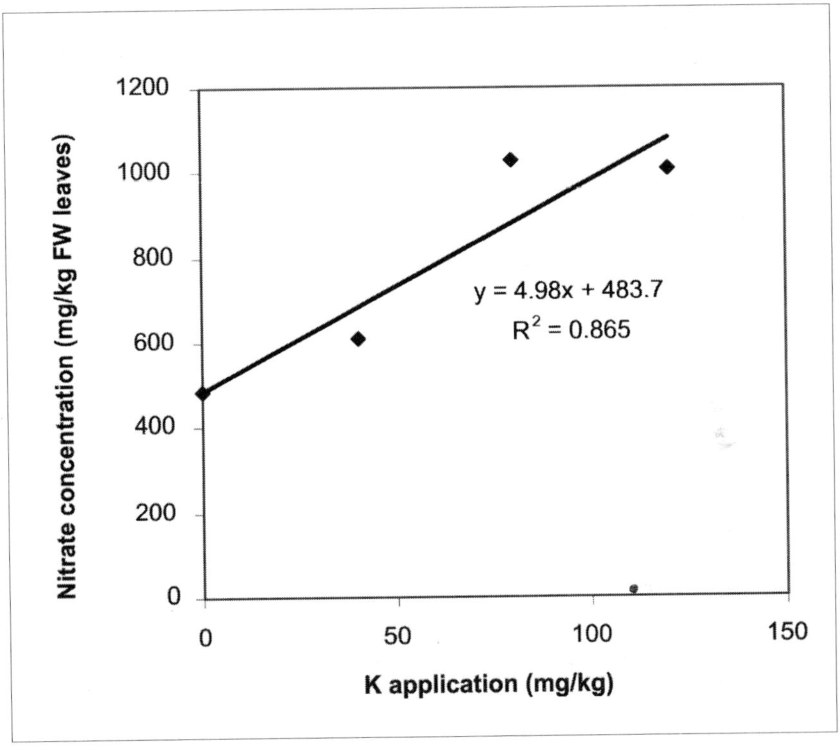

Fig. 7.1. Nitrate concentration in leaves of spinach with increasing application of K.

Table 7.3: Effect of applied potassium on the activities of nitrogen metabolizing enzymes in spinach leaves at 3-week stage of plant growth.

Enzyme activity	Applied K (mg/kg)				**CD (5%)**
	K_0	K_{40}	K_{80}	K_{120}	
Nitrate reductase activity (NRA; ìmoles nitrite h^{-1} mg^{-1} protein)	0.296	0.376	0.597	0.589	0.009
Nitrite reductase (NiR; ìmoles nitrite h^{-1} mg^{-1} protein)	3.210	4.914	5.251	5.217	0.023
Glutamine synthetase (GS; ìmoles GHA min^{-1} mg^{-1} protein)	0.1691	0.2015	0.2905	0.2893	0.004
Glutamate synthase (GOGAT; ìmoles NADH min^{-1} mg^{-1} protein)	0.0332	0.0496	0.0534	0.0522	0.001
Glutamine dehydrogenase (GDH; ìmoles NADH min^{-1} mg^{-1} protein)	0.1928	0.2293	0.2719	0.2695	0.003

Source: Anjana et al. (2009)

NITROGEN-USE EFFICIENCY AND NITRATE REDUCTASE

Nitrogen-use efficiency is a complex term comprising many components. In addition, a great degree of compensation takes place among these components. To measure or quantify nitrogen-use efficiency, the term most widely used is the ratio that considers an output (biological yield or economic yield) as the numerator and input (nitrogen-supply) as the denominator. The biological yield can include either total aboveground plant dry matter or total plant nitrogen, whereas the economic yield includes either grain yield or total grain nitrogen. The nitrogen supply can be from soil, fertilizer (organic or inorganic), or soil plus fertilizer. The ratio of yield to nitrogen-supply is commonly referred to as "agronomic efficiency of nitrogen", the ratio of plant nitrogen to nitrogen-supply is referred to as "recovery efficiency of nitrogen" and the ratio of yield to plant nitrogen is referred to as "physiological or internal efficiency of nitrogen" (Novoa and Loomis 1981). Crop physiological nitrogen-requirements are controlled by the efficiency with which nitrogen in the plant is converted to biomass and grain yield. Therefore, nitrogen use efficiency at the plant level is the ability of the plant to utilize the available nitrogen resources to optimize its productivity (Raghuram et al. 2006).

Intrinsic nitrogen-use efficiency of the whole plant (biomass production per unit nitrogen accumulated) is determined by the integrated biochemical processes in the plant. This includes nitrogen uptake and assimilatory processes, redistribution within the cell and balance between storage and current use at the cellular and the whole plant level. Moreover, since nitrogen-demand and its actual availability tend to vary in time, space and environmental conditions, the regulation of plant nitrogen metabolism must be responsive to nutritional, metabolic and environmental cues (Raghuram et al. 2006). Thus, the efficiency of the processes is genetically determined, though the capacity for biomass production and nitrogen-uptake may be very much dependent on environmental conditions, including nitrate supply (Lawlor 2002).

Different workers have suggested different factors governing the physiological nitrogen-use efficiency: According to Cassman et al. (2002), crop-physiological nitrogen-efficiency is largely governed by two factors: (a) the genetically determined mode of photosynthesis, either the C_3 or C_4 photosynthetic pathway; and (b) the grain-nitrogen concentration, also under genetic control but affected by nitrogen-supply as well. The C_4 plants tend to have a greater physiological nitrogen-efficiency than C_3 plants because the C_4 pathway has a higher photosynthetic rate per unit leaf-nitrogen content, which results in greater biomass production per unit of plant-nitrogen accumulation. However, Cassman et al. (2002) have suggested that there is little scope for genetic improvement in physiological nitrogen-efficiency because the relationship between economic yield and crop-nitrogen uptake is tightly conserved. This in turn suggests only marginal gains in nitrogen-efficiency from molecular engineering of nitrogen-assimilation and biochemical transformation pathways within the plant. Likewise, nitrogen-uptake capacity of crop root systems does not appear to be a sensitive factor limiting the efficiency with which most crops acquire soil or fertilizer-nitrogen, especially when compared to potential improvements in nitrogen-use efficiency from better crop or soil management practices. Therefore, the amount of nitrogen available to the plant can be improved by using sustained release fertilizers, split applications, minimizing fertilizer losses and other nutrient management and crop management strategies. However, the inherent efficiency of the plant to utilize available nitrogen for higher productivity needs to be tackled biologically (Abrol et al. 1999; Abdin et al. 2005).

Moreover, since nitrate reductase (NR) activity is the most limiting step in nitrogen assimilation, attempts have also been made to select genotypes with high NR activity with the view that high NR genotype shall also possess a high nitrogen utilization efficiency (NutE). However, it was found that high NR genotype did not differ much from low NR genotype in NutE due to decline in nitrate-assimilating capacity at later stages of plant growth which in turn is due to a parallel decrease in soil-nitrate content (Abrol, 1990; Anjana et al. 2011, Umar et al. 2011). On this basis, efforts have been made to enhance NutE by individually overexpressing some of the proteins and enzymes responsible for uptake and assimilation of nitrate in transgenic plants, but failed. It indicates that efficient utilization of available nitrate requires a coordinated gene expression and that the earlier notion of a single point rate-limiting regulation was too simplistic. Therefore, there is a need to understand the process in an integrated manner for which natural variability in genotypes for NR activity may be exploited under non-limiting nitrogen supply.

STRATEGIES TO ENHANCE NITROGEN-USE EFFICIENCY

Crop plants have been developed over the last 10,000 years and for most of the time they were not heavily fertilized. However, in the last 50 years, nitrogen fertilization of crop plants has increased worldwide more than 20-fold. The use of fertilizer is generally inefficient with only about 50% being recovered in the harvested crop. Crop plants did not evolve under conditions of high nitrogen nutrition. The question therefore arises whether, based on our knowledge and the experimental techniques now available to us, we can improve the efficiency of nitrogen use by crop plants. We can possibly do it in two ways: (a) by making the best use of natural variability in nitrogen use characteristics available within the gene pool and, (b) by introducing new genes which might increase that variation (Miflin and Habash 2002).

Selection of Genotypes

Plant genotypes with a greater nitrogen-use efficiency, either in a strict physiological sense such as increased carbon gain per unit plant nitrogen per unit time, or in agronomic sense such as greater dry matter or protein yield per unit plant-nitrogen, or per unit nitrogen applied/available to the crop, need to be identified/developed (Anjana et al. 2007c). Greater nitrogen-use efficiency could allow a reduction in nitrogen-fertilizer use without a decrease in yield. However, nitrogen-inputs are inevitable for most crops due to the removal of nitrogen through harvested plant products and hence depletion of soil nitrogen. The approaches to select genotypes with a greater ability to utilize applied or soil available nitrogen and/or nitrogen taken up by the plant have ranged from the primarily agronomic, genotype comparison studies of yield response to applied-nitrogen, to the genetic manipulation of plant characteristics linked to the utilization of nitrogen. The nitrogen-uptake and -utilization varies with crop species and cultivars and within the species of various crops (Pollmer et al. 1979; Reed et al. 1980; Traore and Maranville 1999). Therefore, utilization of plant species or genotypes of same species efficient in absorption and utilization of nitrogen is an important strategy in improving nitrogen-use efficiency and sustainable agricultural systems.

Quantitative Studies of Genetic Variability

Quantitative studies of genetic variability for nitrogen-use efficiency using molecular markers and combining agronomic and physiological studies have been used to identify new genes or loci involved in the regulation of nitrogen metabolic pathway and its interconnection with carbon assimilation and recycling and to select genotypes that assimilate or remobilize nitrogen more efficiently. Thus, it is possible to find genetic variation and select for increased or decreased activities of several enzymes involved in the nitrogen-assimilatory pathway by quantitative genetics (Degenhart et al. 1992; Harrison et al. 2000; Limami and de Vienne 2001). To this end, Harrison et al. (2004) found that the genetic variability in shoot nitrate content was mainly due to an increase in the uptake of the ion, regardless of the plant biomass production in *Lotus japonicus*. It indicates that nitrate transport in *Lotus* is one of the main checkpoints controlling shoot nitrate accumulation. The positive correlation observed between the shoot nitrate content and the steady state level of mRNA encoding high affinity nitrate transporters suggests that the higher nitrate influx is due to enhanced expression of the transporters. In contrast, neither the level of nitrate reductase mRNA, nor the potential enzyme activity in vivo in the different lines was correlated with the shoot nitrate content. Therefore, the current agronomic concept that crop nitrogen-accumulation under suboptimal conditions is highly related to crop growth rate and biomass accumulation via internal plant regulation (Gastal and Lemaire 2002) needs to be reconsidered. It is now necessary to take into account that the control of nitrate uptake and its accumulation in the plant may be subjected to genetic variability regardless of the plant demand. This variability allows for an adaptive regulatory control mechanism depending on the soil nitrate availability (Harrison et al. 2004). Therefore, it is possible through breeding strategies to lower the nitrate content without affecting biomass production.

Molecular Approach

Currently, much effort and enthusiasm is directed to understanding the genomes of plants; hopes of increased food production are focused on genetic modification to increase genetic potential. Genetic manipulation techniques, in which isolated genes are introduced into crops, alter the systems at specific points identified as key, or limiting steps in metabolism and offer the potential for the engineering of metabolic pathways. However, not a single gene for nitrogen-use efficiency or for grain yield or quality, but many genes coding for proteins determine the structures and behaviors of different biochemical and physiological subsystems. It is therefore likely that changes to the genome do not result in major alterations to basic metabolism (Paul and Lawlor 2000). Improvement of crop potential requires identification of specific parts of the metabolism, which must be altered to achieve the given outputs. Genetic characteristics may be specifically altered to increase production and its efficiency, but the targets for modification must be clearly identified (Lawlor 2002).

Over-expression of Nitrate Reducatse

Nitrate reductase (NR), being the first reducing site in the nitrate assimilation pathway, might be the target of choice for genetic manipulation as it possesses negative relationship with nitrate content in leafy vegetables (Anjana et al. 2007a). In an earlier hypothesis, Hageman and his group (1990) suggested a relationship between activity of the enzyme, NR, and grain yield and various other characteristics were suggested. It was postulated that since the enzyme was rate-limiting, any increase in its content/activity will be beneficial and improve nitrogen utilization. Efforts have been made to optimize the nitrogen use efficiency by modulating NR expression pattern. However, as expected for such a highly regulated enzyme, multiple effects were found after overexpression/under expression of NR-genes.

Transgenic plants expressing NR activity at a level five to six times of the wild type obtained by transformation of the E23 *nia* mutant with tobacco *nia* 2 cDNA under the control of the 35S promoter, showed no differences in chlorophyll, proteins or amino acid accumulation. Similarly, an overexpression mediated by light after fusing the NIA1*2 of *Arabidopsis* to LHCB1*13 led to an increase of protein content by 200 % while not displaying advantage with respect to fresh and dry weight (Nejidat et al. 1997). On the other hand, *A. thaliana* mutants affected in *nia* 2 gene, and barley mutants which express only 10 percent of the wild type NR activity, did not show any decline in the nitrogen content and biomass under greenhouse growth conditions (Hoff et al. 1994). Therefore, in overexpressing genotypes nitrate assimilation was not significantly increased due to feedback regulation of nitrate uptake, probably to reduce the excessive amino acid production. In underexpressing genotypes, again reduced nitrate assimilation due to decreased nitrate uptake was discovered indicating that other than reduced nitrogen-compounds are regulatory factors (Gojon et al. 1998). In general, transgenic studies from several plants (reviewed by Andrews et al., 2004) clearly indicate that these enzymes of primary nitrate assimilation are not suitable targets for metabolic engineering to improve nitrogen-use efficiency. Interestingly enough, it was observed that a high NR activity in *Nicotiana* resulted in low nitrate content, compared to the wild ones (Quillere et al. 1994). Similar relationship was observed between the NR activity and nitrate content, which reportedly show two- to threefold variations in wheat cultivars. There was a 30 to 56 percent reduction in nitrate content of high NR cultivars, compared to low NR ones (Ramraj et al. 1999). Therefore, these findings have a bearing on the improvement in the nutritional quality of the leafy vegetables, and an overexpression resulting in a decreased level of foliar nitrate has a commercial importance (Curtis et al. 1999).

Overexpression of Glutamine Synthetase/glutamate Synthase

The cooperative activity of two enzymes, namely, glutamine synthetase (GS) and glutamate synthase (GOGAT) facilitates the ammonia assimilation in most plants. Where, glutamine synthetase catalyzes the first step of ammonium incorporation into glutamate using ATP to yield glutamine (Suzuki and Knaff 2005). GOGAT catalyses the transfer of the amide group of glutamine to 2-oxoglutarate, resulting in the formation of two molecules of glutamate. GS/GOGAT pathway is of crucial importance since the glutamine and glutamate produced are donors of amino groups for the biosynthesis of major nitrogen-containing compounds, including amino acids, nucleotides, chlorophylls, polyamines, and alkaloids (Hirel and Lea 2001; Lea and Ireland 1999).

Considering the aforesaid facts and based on the quantitative studies of genetic variability, several workers have proposed that manipulation of GS_1 could lead to increased nitrogen-use efficiency (Hirel et al. 2001; Miflin and Habash 2002). It was suggested that using transgenic technology, or the other methods, to elevate GS activity in crops offers a promising approach to reducing agricultural nitrogen-fertilization and the concurrent deleterious impacts on the environment. Several groups have attempted to improve nitrogen assimilation by the over expression of GS genes with mixed results (Hemon et al. 1990; Hirel et al. 1992; Temple et al. 1993; Vincent et al. 1997; Miggie et al. 2000; Ortega et al. 2001). It has been demonstrated (Fuentes et al. 2001; Oliveira et al. 2002) that constitutive overexpression of cytosolic GS_1 increases photosynthetic productivity under nitrogen-deficiency, whereas increased growth with increased leaf GS_1 activity is related to increased efficiency in the use of internal nitrogen. The effect of overexpression of leaf GS_2 on plant growth has been tested for tobacco. A twofold increase in GS_2 activity in leaves stimulated the growth of seedlings (Migge et al. 2000). However, a review of this work indicated that although GS activity was correlated with a decrease in the ammonia pool and increases in glutamine and glutamate levels, positive effects on growth were inconsistent (Harrison et al. 2000). The review of the work done on genetic manipulation of GS/GOGAT pathway, has suggested (Andrews et al. 2004) that manipulation of GS/GOGAT linked to primary nitrogen-assimilation would not increase yields. There is little indication that increase in GS/GOGAT linked to primary nitrogen-assimilation would be a useful strategy to increase nitrogen use efficiency.

On the basis of the above findings, it may be concluded that efforts to enhance nitrogen-use efficiency by individually overexpressing some of the proteins and enzymes responsible for the uptake and assimilation of nitrate in transgenic plants have failed, indicating that the earlier notions of a single-point rate-limiting regulation were too simplistic (Stitt 1999; Andrews et al. 2004). Efficient utilization of available nitrate requires a coordinated gene expression and/or post-translational regulation of the proteins/enzymes involved in nitrate transport and reduction, as well as those involved in carbon partitioning for amino acid synthesis. In terms of finding a global target for manipulation of nitrogen-use efficiency, the success in improving nitrate assimilation by overexpressing Dof1 transcription factor (Yanagisawa et al. 2004) indicates that signaling processes provide an attractive route for metabolic engineering (Raghuram et al. 2006).

CONCLUSION AND PERSPECTIVES

Excess input of nitrogen more than its demand leads no or least nitrogen absorption by most of plants in addition to nitrogen buildup in the soil, mostly as nitrates. Leafy vegetables are an important element of human diet. They are an excellent source of vitamins, minerals, sugars, folic acid and they have not much calories. However, leafy vegetables grown under different agro-ecological conditions accumulate nitrate to potentially harmful concentrations. Hence, the greatest exposure to nitrate occurs from the consumption of leafy green vegetables; thus, reigning severe human health concerns. In this context, the potential strategies, namely, the use selection of nitrogen-efficient crops and/or adoption of strategies for enhancing nitrogen-use efficiency may largely work in isolation or combination in reducing ingestion of crop products with high nitrate-mediated anomalies in humans. The two factors, namely, the genetically determined mode of photosynthesis and the grain-nitrogen concentration largely govern the crop-physiological nitrogen-efficiency. However, molecular engineering of nitrogen-assimilation and biochemical transformation pathways within the plant has yielded only marginal gains in nitrogen-efficiency. Nevertheless, nitrogen-uptake capacity of crop root systems does not appear to be a sensitive factor limiting the efficiency. However, the use of sustained release fertilizers, split applications, minimizing fertilizer losses and other nutrient management and crop management strategies can largely improve the amount of nitrogen available to the plant. Moreover, since nitrate reductase activity largely modulates nitrogen assimilation, attempts are being made to select genotypes with high nitrate reductase activity with the view that high nitrate reductase genotype shall also possess a high nitrogen utilization efficiency. In this perspective, a better understanding of the regulatory mechanisms controlling different metabolic processes, i.e., nitrate uptake, nitrate partitioning between roots and shoots, nitrate reduction and its subsequent assimilation and transfer into organic molecules will allow selection of plant criteria and physiological traits that can then be used as markers for improving nitrogen use efficiency in plants.

ACKNOWLEDGEMENTS

SU, Anjana and NAK would like to acknowledge the receipt of funds from DBT, DST, and UGC, CSIR, Govt. of India, New Delhi. NAA is grateful to the Portuguese Foundation for Science and Technology (FCT) for financial support (SFRH/BPD/64690/2009; SFRH/BPD/84671/2012).

REFERENCES

Abdin M.A., Dwivedi R.K. and Abrol Y.P. (2005). Nitrogen in agriculture, *In*: Singh, R.P., Shankar, N. and Jaiwal, P.K. (Eds.). *Nitrogen nutrition and plant productivity*. Studium Press, Texas, pp. 1-41.

Abrol, Y. P. (1990). Pattern of nitrate assimilation and grain nitrogen yield in field-grown wheat (*Triticum aestivum*). *In*: M.L.V. Beusichem *(Ed.). Plant Nutrition-Physiology and Applications*. Kluwer, Dodrecht, the Netherlands, pp. 773–778.

Abrol, Y.P., Chatterjee, W.R., Kumar, P.A. and Jain, V. (1999). Improvement in nitrogen use efficiency: Physiological and molecular approaches. *Curr. Sci.* **76:** 1357-1364.

Andrews, M., Lea, P.J., Raven, J.A. and Lindsey, K. (2004). Can genetic manipulation of plant nitrogen assimilation enzymes result in increased crop yield and greater N-use efficiency? An assessment. *Ann. Appl. Biol.* **45:** 25-40.

Anjana, Umar, S., Iqbal, M. and Abrol, Y. P. (2007a). Are nitrate concentrations in leafy vegetables within safe limits? *Curr. Sci.* **92** (3)**:** 355-360.

Anjana, Umar, S. and Iqbal, M. (2007b). Nitrate accumulation in plants, factors affecting the process, and human health implications - a review. *Agron. Sustain. Dev.* **27:** 45-57.

Anjana, Umar, S. and Iqbal, M. (2007c). Identification of characteristics affecting nitrogen utilization efficiencies in wheat cultivars. *Arch. Agron. Soil Sci.* **53 (4):** 459-472.

Anjana, Umar, S. and Iqbal, M. (2009). Effect of Applied Potassium in Increasing the potential for Nitrogen Assimilation in Spinach (*Spinacia oleracea* L.). *e-ifc* **20**:8-10.

Anjana, Umar, S., Abrol, Y. P. and Iqbal, M. (2011). Modulation of nitrogen utilization efficiency in wheat genotypes differing in nitrate reductase activity *J. Plamt Nutr.* **34:** 920-933.

Caldwell, C.R. and Britz, S.J. (2006). Effect of supplemental ultraviolet radiation on the carotenoid and chlorophyll composition of green house-grown leaf lettuce (*Lactuca sativa* L.) cultivars. *J. Food Comp. Anal.* **19:** 637-644.

Cardenas-Navarro, R., Adamowicz, S. and Robin, P. (1999). Nitrate accumulation in plants: a role for water. *J. Exp. Bot.* **50:** 613-624.

Carranca, C., Soares da Silva, A., Fernandes, M. and Varela, J. (2001). 15N fertilizer use efficiency by spinach grown under portuguese field conditions. *Acta Hort.* **563:** 67-72

Cassman, K.G., Dobermann, A. and Walters, D.T. (2002). Agroecosystems, nitrogen-use efficiency, and nitrogen management. *Ambio* **31:** 132-140.

Curtis, I.S., Power, J.B., deLaat, A.M.M., Caboche, M. and Davey, M.R. (1999). Expression of a *chimeric nitrate reductase* gene in transgenic lettuce reduces nitrate in leaves. *Plant Cell Rep.* **18:** 889-896.

Degenhart, N.R., Barnes, D.K. and Vance, C.P. (1992). Divergent selection for nodule aspartate aminotransferase and asparagines synthetase activities in alfalfa. *Crop Sci.* **32:** 313-317.

Dich, J., Jrvinen, R., Knekt, P. and Penttil, P.L. (1996). Dietary intakes of nitrate, nitrite and NDMA in the finish mobile clinic health examination survey. *Food Add. Cont.* **13:** 541-552.

Drescher, A., Glaser, R., Richert, C. and Nippes, K.R. (2011). Demand for key nutrients (NPK) in the year 2050. Draft report. Department of Geography, University of Freiburg.

Durner, J. and Klessig, D.F. (1999). Nitric oxide as a signal in plants. *Curr. Opin. Plant Biol.* **2:** 369-374.

E.F.S.A. Panel on Contaminants in the Food Chain (CONTAM) (2010). Scientific Opinion on possible health risks for infants and young children from the presence of nitrates in leafy vegetables. *EFSA Journal* **8(12):**1935.

F.A.O. (2010a). FAO: the State of World Food Insecurity in the World.

F.A.O. (2010b). FAO: climate Smart Agriculture. Policies, Practices and Financing for Food Security, Adaptation and Mitigation.

Ferrario-Mery, S., Murchie, E., Hirel, B., Galtier, N., Quick, W.P. and Foyer C.H. (1997). Manipulation of the pathways of sucrose biosynthesis and nitrogen assimilation in transformed plants to improve photosynthesis and productivity. *In:* Foyer, C.H. and Quick, W.P. (eds.). *A molecular approach to primary metabolism in higher plants.* Taylor and Francis, London, pp. 125-153.

Fuentes, S.I., Allen, D.J., Ortiz-Lopez, A. and Hernandez, G. (2001). Over-expression of cytosolic gultamine synthetase increases photosynthesis and growth at low nitrogen concentrations. *J. Exp. Bot.* **52:** 1071-1081.

Gastal, F. and Lemaire, G. (2002). N uptake and distribution in crops: an agronomical and ecophysiological perspective. *J. Exp. Bot.* **53:** 789-799.

Gojon. A., Dapoigny, L., Lejay, L., Tillard, P. and Rufty, T.W. (1998). Effects of genetic modification of nitrate reductase expression of $^{15}NO_3^-$ uptake and reduction in *Nicotiana* plants. *Plant Cell Environ.* **21:** 43-53.

Gupta, S.K., Gupta, R.C., Gupta, A.B., Seth, A.K., Bassin, J.K. and Gupta, A. (2001). Recurrent diarrhea in areas with high nitrate in drinking water. *Arch. Environ. Health* **56:** 369-374.

Gupta, S.K., Gupta, R.C., Seth, A.K., Gupta, A.B., Bassin, J.K., Gupta, D.K. and Sharma, S. (1999). Epidemiological evaluation of recurrent stomatitis, nitrates in drinking water and cytochrome b5 reductase activity. *Am. J. Gastroenterol.* **94:** 1808-1812.

Hageman, R.H. and Below, F.E. (1990). Role of nitrogen in crop productivity. *In:* Abrol, Y.P. (Ed.). *Nitrogen in higher plants.* Wiley and Sons, New York, pp. 313-334.

Hakeem, K.R., Chandna, R., Ahmad, A. and Iqbal, M. (2012). Reactive nitrogen inflows and nitrogen use efficiency in agriculture: an environment perspective. *In:* Ahmad, P. and Prasad, M.N.V. (eds.). *Environmental adaptations and stress tolerance of plants in the era of climate change.* Spinger, The Netherlands. pp 217-232.

Harrison, J., Brugiere, N., Phillipson, B., Ferrario-Mery, S., Becker, T., Limami, A. and Hirel, B. (2000). Manipulating the pathway of ammonia assimilation through genetic manipulation and breeding. Consequences on plant physiology and development. *Plant Soil* **221:** 81-93.

Harrison, J., Hirel, B. and Limani, A.M. (2004). Variation in nitrate uptake and assimilation between two ecotypes of *Lotus japonicus* and their recombinant inbred lines. *Physiol. Plant* **120:** 124-131.

Hemon, P., Robbins, M. and Cullimore, J. (1990). Targeting of glutamine synthetase to the mitochondria of transgenic tobacco. *Plant Mol. Biol.* **15:** 895-904.

Hirel, B., Bertin, P. and Quillere, I. (2001). Towards a better understanding of the genetic and physiological basis for nitrogen use efficiency in maize. *Plant Physiol.* **125:** 1258-1270.

Hirel, B., Marsolier, M., Hoarau, A., Brangeon, J., Schafer, R. and Verma, D.P.S. (1992). Forcing expression of a soybean root glutamine synthetase gene in tobacco leaves induces a native gene encoding cytosolic enzyme. *Plant Mol. Biol.* 20: 207-218.

Hoff, T., Truong, H.-N. and Caboche, M. (1994). The use of mutants and transgenic plants to study nitrate assimilation. *Plant Cell Environ.* **17:** 489-506.

Ikemoto, Y., Teraguchi M. and Kogayashi, Y. (2002). Plasma level of nitrate in congenital heart disease: Comparison with healthy children. *Pediatr. Cardiol.* **23:** 132-136.

Ishiwata, H., Yamada, T., Yoshiike, N., Nishijima, M., Kawamoto, A. and Uyama, Y. (2002). Daily intake of food additives in Japan in five age groups estimated by the market basket method. *Eur. Food Res. Technol.* **215:** 367-374

http://www.livestrong.com/article/446299-what-vegetables-have-nitrates/ (Accessed on 16 Nov. 2012).

Ko³ton, A. and Baran, A. (2008). Effect of different mineral nitrogen and compost nutrition on some compounds of corn salad (*Valerianella locusta* (L.) Latter.). *Sodininkyste Ir Dar•ininkyste* **27(2):** 379-386.

Lamattina, L., Garcia-Mata, C., Graziano, M. and Pagnussat, G. (2003). Nitric oxide: The versatility of an extensive signal molecule. *Annu. Rev. Plant Biol.* **54:** 109-136.

Larson, R.A. (1988). The antioxidants of higher plants. *Phytochemistry* **27:** 969-978.

Lawlor, D.W. (2002). Carbon and nitrogen assimilation in relation to yield: mechanisms are the key to understanding production systems. *J. Exp. Bot.* **53:** 773-787.

Lea, P.J. and Ireland, R.J. (1999). Nitrogen metabolism in higher plants. *In:* Singh, B.K. (ed.). *Plant amino acids. Biochemistry and biotechnology.* Marcel Dekker, New York, United States, pp.1-47.

Limami, M.A. and de Vienne, D. (2001). Natural genetic variability in nitrogen metabolism. *In:* Lea, P.J. and Morot-Gaudry, J.F. (Eds.). *Plant nitrogen*. Springer-Verlag, Berlin, pp. 368-397.

Ma, L. and Dolphin, D. (1999). The metabolites of dietary chlorophylls. *Phytochemistry* **50:** 195-202.

McIntyre, G.I. (1997). The role of nitrate in the osmotic and nutritional control of plant development. *Aust. J. Plant Physiol.* **24:** 103-118.

Miflin, B.J. and Habash, D.J. (2002). The role of glutamine synthetase and glutamate dehydrogenase in nitrogen assimilation and possibilities for improvement in nitrogen utilization of crops. *J. Exp. Bot.* **53:** 979-987.

Migge, A., Bork, C., Hell, R. and Becker, T.W. (2000). Negative regulation of nitrate reductase gene expression by glutamine or asparagines accumulating in leaves of sulfur-deprived tobacco. *Planta* **211:** 587-595.

N.A.A.S. (2005). Policy options for efficient nitrogen use. Policy paper No. 33, National Academy of Agricultural Sciences, New Delhi. pp. 1-4.

Nejidat, A., Zhang, G., Grinberg, M. and Heimer, Y.M. (1997). Increased protein content in transgenic *Arabidopsis thaliana* over-expressing nitrate reductase activity. *Plant Sci.* **130:** 41-49.

Nosengo, N. (2003). Fertilized to death. *Nature* **425:** 894-895.

Novoa R. and Loomis, R.S. (1981). Nitrogen and plant production. *Plant Soil* **58:** 177-204.

Oliveira, I.C., Brears, T., Knight, T.J., Clark, A. and Corruzi, G.M. (2002). Overexpression of cytosolic glutamine synthetase. Relation to nitrogen, light, and photorespiration. *Plant Physiol.* **129:** 1170-1180.

Ortega, J.L., Temple, S.J. and Sengupta-Gopalan, C. (2001). Constitutive overexpression of cytosolic glutamine synthetase (GS_1) may be regulated at the level of RNA stability and protein turnover. *Plant Physiol.* **126:** 109-121.

Ourry, A., Gordon, A.J. and Macduff, J.H. (1997). Nitrogen uptake and assimilation in roots and root nodules. *In:* Foyer, C.H. and Quick, W.P. (eds.). *A molecular approach to primary metabolism in higher plants.* Taylor and Francis, London, pp. 237-253.

Paul, M.J. and Lawlor, D.W. (2000). Genetic manipulation of photosynthesis. *J. Exp. Bot.* **51:** 319-496.

Pollmer, W.G., Eherhard, D., Klein, D. and Dhillon, B.S. (1979). Genetic control of nitrogen uptake and translocation ion maize. *Crop Sci.* **19:** 82-86.

Prakasa Rao, E.V.S. and Puttanna, K. (2000). Nitrates, agriculture and environment. *Curr. Sci.* **79:** 1163-1168.

Quillere, I., Dufosse, C., Roux, Y., Foyer, C.H., Caboche, M. and Morot-Gaudry, J.F. (1994). The effects of deregulation of NR gene expression on growth and nitrogen metabolism of *Nicotiana plumbaginifolia* plants. *J. Exp. Bot.* **45:** 1205-1211.

Raghuram, N., Pathak, R.R. and Sharma, P. (2006). Signalling and the molecular aspects of N-use-efficiency in higher plants. *In:* Singh, R.P. and Jaiwal, P.K. (eds.). *Biotechnological approaches to improve nitrogen use efficiency in plants.* Studium Press, Texas, pp. 19-40.

Ramraj, V.M., Guru, S.K. and Abrol, Y.P. (1999). Nitrate content in wheat leaf blades. *Curr. Sci.* **76:** 29-30.

Reddy, K.S. and Menary, R.C. (1990). Nitrate reductase and nitrate accumulation in relation to nitrate toxicity in *Boronia megastigma*. *Physiol. Plant* **78:** 430-434.

Reed, A.J., Below, F.E. and Hageman, R.H. (1980). Grain protein accumulation and the relationship between leaf nitrate reductase and protease activities during grain development in maize. I. Variation between genotypes. *Plant Physiol.* **66:** 164-170.

Reed, A.J., Below, F.E. and Hageman, R.H. (1980). Grain protein accumulation and the relationship between leaf nitrate reductase and protease activities during grain development in maize. I. Variation between genotypes. *Plant Physiol.* **66:** 164-170.

Sharma, M.K. and Sharma, H. (2012). Evaluation of the haematological responses to high nitrate exposure in rabbits. *J Clin. Diagnos. Res.* **6:** 145-149.

Stitt, M. (1999). Nitrate regulation of metabolism and growth. *Curr. Opin. Plant Biol.* **2:** 178-186.

Suzuki, A. and Knaff, D.B. (2005). Glutamate synthase: structural, mechanistic and regulatory properties, and role in the amino acid metabolism. *Photosynth. Res.* **83:** 191-217.

Temple, S.J., Knight, T.J., Unkefer, P.J. and Sengupta-Gopalan, C. (1993). Modulation of glutamine synthetase gene expression in tobacco by the introduction of an alfalfa glutamine synthetase gene in sense and antisense orientation: Molecular and biochemical analysis. *Mol. Gen. Genet.* **236:** 315-325.

Traore, A. and Maranville, J.W. (1999). Nitrate reductase activity of diverse grain sorghum genotypes and its relationship to nitrogen use efficiency. *Agron. J.* **91:** 863-869.

Umar, S., Anjana and Iqbal, M. (2011). Interactive effects of potassium and nitrogen nutrition on physiological use efficiency of nitrogen and crop yield. *In*: Jain, V. and Kumar, P. A. (eds.). *Nitrogen Use Efficiency in Plants.* New India Pub., pp. 125-155.

U.N. (2011). World Population Prospects: The 2010 Revision. http://esa.un.org/unpd/wpp/index.htm.

Vincent, R., Fraisier, V., Chaillou, S., Limami, M.A., Deleens, E., Phillipson, B., Douat, C., Boutin, J.P. and Hirel, B. (1997). Over expression of a soyabean gene encoding cytosolic glutamine synthetase in shoots of transgenic *Lotus cornicultus* L. plants triggers changes in ammonium and plant development. *Planta* **201:** 424-433.

Vinson, J.A., Hao, Y., Su, X. and Zubik, L. (1998). Phenol antioxidant quantity and quality in foods: vegetables. *J. Agric. Food Chem.* **46:** 3630-3634.

Yanagisawa, S., Akiyama, A., Kisaka, H., Uchimaya, H. and Miwa, T. (2004). Metabolic engineering with Dof1 transcription factor in plants: Improved nitrogen assimilation and growth under low-nitrogen conditions. *Proc. Nat. Acad. Sci. USA* **101:** 7833-7838.

8

Crop, Soil, and Nitrogen Fertilization Management Strategies to Reduce Nitrate Accumulation in Plants

Harmandeep Singh[1], **Kaushik Majumdar**[2] **and T. Satyanarayana**[3]

[1]*International Agronomic and Technical Support Specialist, International Plant Nutrition Institute (IPNI), 102-411 Downey Road, Saskatoon, SK S7N 4L8, Canada*
[2]*IPNI, India and Bangladesh Programs, 354, Sector-21, HUDA, Gurgaon – 122 016, Haryana*
[3]*IPNI, South India, 401, Tejaswini Arcade, Plot No. 78, 10-2-99, 99/2, Opposite More Super Market, West Marredpally, Secunderabad – 500 026, Andhra Pradesh*
e-mail: kmajumdar@ipni.net

ABSTRACT

Balancing the amount of N needed for optimum plant growth while reducing nitrate accumulation in plants and other N losses from soil remains a major challenge for everyone attempting to understand and improve agricultural nutrient use efficiency. The objectives for this chapter are to discuss current crop, soil, and N fertilization management strategies and identify future research needs aimed at reducing nitrate accumulation in plants. Several current and potential N management practices are being evaluated and used to reduce the nitrate loading. The problem is not solely caused by N fertilizer management or by any other single factor, but is a combination of soil management practices and inherent physical, chemical, and biological characteristics of the soil as well as the physiological and environmental factors. Therefore, based on current experiences, a combination of two or more management practices may work in harmony and also may make implementation of the practices more cost effective. The most desirable outcome associated with research and technology development to reduce nitrate accumulation in plants is a better public recognition that several different management practices will be required for this purpose.

Keywords: Bioreactors, Chlorophyll monitoring, Cover crops, Diversified crop rotations, Nitrate accumulation, Nitrate removal methods, Nitrification inhibitors, Nitrogen application rate, Nitrogen application time, Nitrogen fertilization, Nitrogen fertilizer application equipment, Nitrogen mineralization, Nitrogen use efficiency, Plant residue, Riparian buffers, Tillage, Variable-rate nitrogen application, Wetlands

INTRODUCTION

Nitrogen (N) is essential for the growth and reproduction of all life forms, and except for legume crops and virgin soils with relatively high soil organic matter (SOM), soil N must usually be supplemented to sustain food, feed, and fiber production. But in our never ending quest to increase the production, excessive amounts of nitrogenous fertilizers are applied to crops. This is mainly because N deficiency in soils is universal, and applying fertilizer N is considered a reasonable insurance against crop yield losses and their economic consequences. However, when input of N exceeds its demand, plants are no longer able to absorb excess N, which then builds up in the soil, mostly as nitrates (Nosengo 2003). This build up not only causes imbalance of nutrients in the soil, but also increases the nitrate level in groundwater supplies (NAAS 2005), ultimately impacting the nitrate content of plants (Vieira et al. 1998; Dapoigny et al. 2000). Secondly, current practices of N fertilizer management are often very inefficient compared with the natural systems. For example, first-year plant recoveries of fertilizer N have been reported to be 35% (Bijeriego et al. 1979), 24 to 26% (Olson 1980), 23 to 45% (Kitur et al. 1984), 14 to 65% (Meisinger et al. 1985), 15 to 33% (Sanchez and Blackmer 1988), and 45 to 59% (Reddy and Reddy 1993), 33% (Raun and Johnson 1999), 30% (Dobermann 2000). The 35 to 86% unrecovered fertilizer N increases the potential for nitrate build up in soils, and further in plants. Thirdly, some physiological and environmental factors also increase the accumulation of nitrates in plants. For example, in the early period of plant growth, nitrate uptake is relatively high and nitrates may become concentrated. As plant growth slows, the plant continues to take up nitrates, but the nitrates do not get converted to amino acids and proteins, thereby again leading to nitrate concentration. Periods of drought or cloudy weather that slow down the rate of photosynthesis are often associated with slow plant growth and the accumulation of nitrates. Rapid uptake of nitrates during the first few hours following a drought-ending rain may also cause high nitrate levels in the plant. The conversion of nitrates to amino acids within the plant may take several days. Thus, nitrates remain concentrated for seven or more days after the crop has received enough rain to sustain growth. Also, plants in vegetative stages are most likely to have high nitrate levels, and nitrates are more highly concentrated in stems and leaves closer to the soil surface (Hancock 2007).

The increased build up of nitrate (NO_3^-) in plants is harmful for plants, animals, and humans alike. A high nitrate accumulation in plants results in nitrite (NO_2^{2-})

production, which is converted into nitric oxide (NO). This NO, together with oxygen (O_2^-), could be rapidly catalysed by nitrate reductase into peroxynitrite ($ONOO^-$), which is highly toxic to plants (Anjana et al. 2007). In animals, ingestion of high concentration of nitrates builds up nitrite levels, and the nitrite is absorbed into the animal's bloodstream. Once in the bloodstream, nitrites bind with hemoglobin (creating methemoglobin) and prevent the normal transfer of oxygen. Low nitrate intake levels often cause lower milk production, abortion, breeding problems, or symptoms that mimic a nutritional deficiency (Vitamin A and E, rickets, phosphorus, or calcium imbalance). High nitrate intake levels cause bluish color of mucous membranes, dark brown or "chocolate-colored" blood, rapid and difficult breathing, a rapid pulse (≥150 beats/minute), tremors, staggering, collapse, and even death (Hancock 2007). In humans, nitrate again gets reduced to nitrite, which reacts with amines and amides to produce nitrosamines and nitrosamides whose carcinogenic properties are well known (Walker 1990). Also, like in animals, the production of methemoglobin impairs oxygen delivery to human tissues. A high percentage of acute respiratory tract infection with a history of recurrence has been reported in children consuming high nitrate content (Gupta et al. 2000). Some other reported effects are infant mortality, early onset of hypertension, hypothyroidism, diabetes, and an adverse effect on cardiac muscles, alveoli of the lungs, and adrenal glands (Gupta 2006).

The objectives of this chapter are to (i) discuss current crop, soil, and N fertilization management strategies for reducing nitrate accumulation in plants and (ii) identify future research needs to improve some current strategies and to develop new strategies for reducing nitrate accumulation in plants.

CURRENT CROP, SOIL, AND NITROGEN FERTILIZATION MANAGEMENT STRATEGIES

Timing, Rate, and Method of Nitrogen Application

Besides nitrate accumulation in plants, excess N buildup in soil is also susceptible to leaching, denitrification, volatilization, and immobilization processes within the soil environment. The risk of N losses due to these processes increases as the time between N application and crop uptake increases. This is true for residual N as well as applied N (Magdoff 1991; Karlen et al. 1998), especially in years that do not produce optimal yields (Power et al. 1998). The challenge, therefore, is to manipulate N availability before, during, and after peak crop demand to reduce nitrate accumulation and also to reduce N losses from the soil. Limiting the amount of inorganic N within the soil at the end of a crop's growing season and before the next crop has established an extensive root system, which is a key factor for reducing N losses. Therefore, although timing, method of N application, and accounting for mineralization soil N are important for reducing nitrate accumulation and N losses, Power and Schepers (1989) concluded that the most important factor was to apply the correct amount of N fertilizer.

Most conventional methods of N fertilizer recommendation were developed on a scale of a state or a region. However, recent research conducted in different countries (Angus et al. 1990; Ferguson et al. 1995; Kitchen et al. 1995; Vetch et al. 1995; Wopereis et al. 1999; Wang et al. 2001; Dobermann et al. 2002) has demonstrated limitations of the currently widespread blanket fertilizer N recommendations. This might be due to the existence of large variability in soil nutrient supply and crop response to nutrients among, or even within, farms. For example, Dawe and Moya (1999) found that optimal N rates for rice varied widely from year to year in farmer's fields in Philippines. Cassman et al. (1996a,b) observed that indigenous N supply was variable among fields and seasons. Power et al. (2000) reported that Midwestern fields in USA frequently have a high degree of variability in soil nitrate content from site to site within a single field. They stated that "….soils are seldom uniform throughout a field, so applying sufficient fertilizer N to assure high yields for more productive areas of the field often results in over-fertilization of less productive areas. This may lead to greater N losses and nitrate accumulation in plants, particularly in those areas of the field that are more susceptible to leaching." To be effective, variable-rate technology must accurately identify within-field spatial and temporal variability and reliably interpret these patterns of variability (Sawyer 1994). If this is accomplished, a reliable variable-rate or site-specific N fertilizer application program could reduce the application on areas requiring little or no additional N to optimize yield (Dinnes et al. 1998) and could limit application on areas sensitive to N losses (Larson and Robert 1991).

Pierce and Nowak (1999) discussed three basic management approaches currently being tested for variable rate N application. The first involves determining plant available N levels from field grid sampling and interpreting N rates based on current recommendation methods (i.e., a N balance equation). The second approach bases N rates on observed crop N responses using replication strips with varying N rates across the landscape. The third approach involves determining crop N status by monitoring (i.e., light reflectance or chlorophyll content). Usually, this intervention-type approach uses a portion of the crop that is well fertilized as a standard for comparison. The best approach for a particular field, be it conventional or site-specific, will depend on the amount of spatial and temporal N variability. Pan et al. (1997) reported that temporal N variability could frequently exceed spatial variability. While highly predictable spatial variability may be amenable to multiple approaches of site-specific N management, strong temporal variability is much more difficult to manage (Pierce and Nowak 1999). In this case, they concluded that an intervention strategy would be the most appropriate option.

Methodologies to accurately assess within-field spatial and temporal variability for subsequent N application have yet to be fully developed. Therefore, the key under these circumstances is to manipulate the timing of N application for reducing the excessive buildup of soil nitrate. Khurana et al. (2007, 2008) clearly demonstrated that the current N management practices in Punjab, India, were inconsistent with the physiological requirement of rice and wheat crops, thereby leading to low N use efficiencies and more soil nitrate build up. In their study, N supply appeared to be excessive during the early vegetative growth but deficient during grain filling. Therefore, they devised a site-specific N management strategy in which N applications were more uniform among the sites spread more evenly through the growing season, and avoided heavy single applications at early growth stages of wheat. And through this strategy they were able to increase agronomic and recovery efficiencies of N by 50 to 80%. Similarly, Randall et al. (1992) and Randall (1997) demonstrated that changing the timing of a single preplant fertilizer application from fall to spring could significantly decrease N loss and increase nitrogen use efficiency (NUE) by 20% in southern Minnesota.

Crediting Nitrogen Mineralization

Nitrate, as an end product of mineralization and subsequent nitrification of SOM, manure, crop residue, or previously applied fertilizer N that has cycled through soil organic N pools, can make significant contributions toward meeting crop N requirements (Oberle and Keeney 1990; Kanwar et al. 1996). However, to ensure that nitrate is used efficiently, it is important to quantify the amount being produced and, where possible, minimize its accumulation during non-cropping periods. Strategies to slow nitrate accumulation include the use of nitrification inhibitors (Kidwaro and Kephard 1998) or temporary immobilization by cycling N through cover crops (Doran and Smith 1991; Staver and Brinsfield 1998; Wagger et al. 1998). These strategies are discussed in detail in the following sections.

Monitoring N mineralization to better match the required amount of available N with crop needs is one strategy to reduce soil nitrate buildup. To accomplish this task, several versions of a pre-sidedress soil nitrate test (PSNT) (Magdoff et al. 1984; Fox et al. 1989; Magdoff et al. 1990) or modifications such as the late-spring nitrate test (LSNT) (Blackmer et al. 1997) have been developed. These tests generally involve sampling the soil approximately 6 wk after planting. The philosophy behind these tests is that by late spring, the net effects of mineralization and N losses that may have occurred since the last crop was harvested can be accurately assessed. Soil test results can then be used to determine the appropriate amount of additional N fertilizer to apply.

Plot-scale studies using PSNT or LSNT strategies to determine N rates have generally shown reductions in measured or potential soil N losses. In Iowa, these procedures resulted in fertilizer N applications ranging from 50 to 168 kg N/ha and significantly reduced nitrate loss to water compared with single pre-plant applications of only 112 kg N/ha (Kanwar et al. 1996). A study in Vermont (Durieux et al. 1995) also showed reduced nitrate leaching potential compared with a yield-goal N management strategy. These plot-scale results suggest that the PSNT-LSNT approach for N management has the potential for avoiding excess N buildup in soil through avoiding excess N application compared with yield-goal approaches (Magdoff 1991; Durieux et al. 1995; Karlen et al. 1998). However, while designed to provide an optimum N fertilizer rate at an appropriate time to improve crop N use efficiency, the impacts on water quality and risk to the farmer for adopting this approach have been poorly quantified at the whole-field scale and have only begun to be evaluated at the watershed scale. A regional study (Bundy et al. 1999) determined that the PSNT-LSNT program at the watershed scale failed to identify some field sites where a yield response to N fertilizer was not achieved. Their follow-up investigation found that such errors were reduced by increasing the soil sampling depth from 30 to 60 cm. However, the acceptability and practical adaptability of this approach is still in question.

The end-of-season basal stalk nitrate test (Blackmer and Mallarino 1997) has also been used to verify the accuracy of PSNT-LSNT program. Stalk nitrate concentrations are categorized into four groups: low (<250 mg/kg nitrate), marginal (250–700 mg/kg nitrate), optimal (700–2000 mg/kg nitrate), and excess (>2000 mg/kg nitrate). Although the information does not assist in correcting in-season deficiencies, it is useful for guiding future operations to improve NUE.

Fertilizer Application Equipment

In measuring the efficiency of any system, the variation or error of each step is additive. Therefore, the overall efficiency is only as good as the summed margins of error for each step. For site-specific N management, attempting to apply variable N rates will do little to improve overall NUE if the application equipment has a high degree of error. This problem was confronted in studies examining the performance of conventional anhydrous ammonia manifolds. Weber et al. (1993) found appreciable degrees of error over the entire fields, and other investigators documented large degrees of error across the individual outlets of the equipment' manifold (Reichenberger 1994; Fee 1997; Schrock et al. 1999).

Research efforts have concentrated on evaluating new technology to reduce equipment application error. Boyd et al. (2000) conducted an evaluation of anhydrous ammonia manifolds under field conditions at two target application rates (84 and 168 kg N/ha). Several anhydrous ammonia manifold entry methods were used for conventional manifolds, and although each method had differing coefficients of variation at both N rates, all had similar trends, with greater coefficients of variation at 84 kg N/ha (12–80%) than at 168 kg N/ha (10–66%). Alternative manifold designs typically performed better than the conventional manifolds at both rates, with less variability at the high rate. Rotaflow and small housing Vertical Dam manifolds produced the least amount of variability across all outlets at both the low and high N rates. A John Blue FD-1200 manifold had a low variability at the higher rate.

Liquid N fertilizer applicators tend to have less variation in fertilizer distribution across all outlets along the implements' toolbars than the conventional anhydrous ammonia applicators. Recently, liquid N fertilizer applicators have become even more accurate with the use of hydraulic flow rate control devices. With careful installation and calibration, applied rates can easily be within 1 to 4% of the target rate (Dinnes et al. 2002).

Some recent innovations in the design and function of N fertilizer application equipment have focused on strategies to limit N losses during application. A localized compaction and doming (LCD) applicator (Ressler et al. 1997), was developed to alter soil physical properties immediately above the soil volume where knife-injected liquid N fertilizer was placed. A compacted zone with a higher bulk density is created above the injected zone so that water infiltration through that area would be reduced compared with a conventional knife-injection applicator. When compared with either conventional knife or broadcast methods of N application, the LCD design reduced the amount of N application loss.

Another application technique developed to improve N management was the point injector, which, until no-till conditions, was demonstrated to have the potential to reduce ammonia volatilization and N immobilization at the surface without destroying surface residues or adversely affecting corn yield (Baker et al. 1985). Agricultural drainage studies in Iowa from 1995 through 1997 (Iowa Dept. of Agric. and Land Stewardship 1997) showed that both LCD and point injection methods consistently reduced nitrate concentration in tile drainage water compared with conventional knife-injection methods, with LCD also resulting in the highest corn grain yield.

Nitrification Inhibitors

Nitrification inhibitors (NIs) for N fertilizer have met with varying success, generally depending on the soil type and weather pattern under which they were used. These inhibitors function by limiting the activity and population of *Nitrosomonas* bacteria that convert ammonical ion to nitrogen dioxide. The primary use of NIs is to slow the conversion of ammonia-based fertilizers to nitrate form, thus potentially reducing soil nitrate buildup.

An 8-yr project in Ohio (Stehouwer and Johnson 1990) examined different application timings with and without a nitrification inhibitor. The results showed that at similar N rates, spring preplant urea application produced higher yields than fall applications. Addition of nitrapyrin as a NI with spring-applied N did not increase the yields, but did increase yields with fall N applications. Similarly, in Minnesota, Randall et al. (1992) found that corn yield and NUE were lowest with fall N application without nitrapyrin, highest with spring N applications, and intermediate with fall N plus nitrapyrin. They reported that NUE of fall-applied N fertilizer was 16%, whereas with nitrapyrin, it improved to 26%. Spring applications, with or without nitrapyrin, had fertilizer NUEs ranging from 42 to 48%. Abbasi et al. (2003) demonstrated in their study in Kashmir, India, a 30–40% increase in the apparent recovery of added N with nitrapyrin. Majumdar (2002) showed a tree product *Karanjin* to be a powerful NI that increased NUE and reduced N losses remarkably when compared with an inorganic NI dicyandiamide (DCD).

For widespread adoption of NIs, favorable economics for using NIs is a must. The economics of nitrapyrin use have shown mixed results. In a 7-yr study by Christensen and Huffman (1992), nitrapyrin in spring-applied N fertilizer significantly increased corn grain yield, with the increase more than compensating for the added cost of nitrapyrin. In Minnesota, 7-yr averages of corn fertilized with 150 kg N/ha showed that nitrapyrin did not result in an economic advantage for the spring and split N application treatments, but there was economic advantage for the fall-applied N. However, a number of other studies have found no economic advantage of using nitrapyrin (Hendrickson et al. 1978; Touchton et al. 1979; Blackmer and Sanchez 1988). The variable results with nitrapyrin use are probably due to differences among the factors affecting the microbial process of nitrification (e.g., climate and soil type, etc.).

Chlorophyll Monitoring

Chlorophyll monitoring is done to guide in-season fertilizer N applications with an objective of improving overall NUE. In some studies, chlorophyll measurements have been shown to correlate well with N concentrations in the plant tissue and to have the ability to predict grain yield (Wood et al. 1992; Siambi et al. 1999). In Pennsylvania, Piekielek and Fox (1992) showed that the chlorophyll meter measures were as accurate as several soil N availability tests in identifying N responsive and nonresponsive sites. However, the chlorophyll meter did not correlate well enough with soil N-supplying capacity (in-season mineralization) to accurately make N sidedress recommendations. Bijay-Singh et al. (2002) showed that plant need-based N management through chlorophyll meter reduces N requirement of rice in Northwest India from 12.5 to 25% with no loss in yield. Chlorophyll meter in combination with a sufficiency index was very successful as an in-season N management strategy for irrigated corn in Nebraska (Varvel et al. 1997).

Currently, it is unlikely that chlorophyll meter data alone will be sufficient to guide N fertilizer applications. Factors other than N availability affect leaf chlorophyll content like plant leaf chlorosis due to nutrient deficiencies other than N (e.g., K and Mg), disease, insect damage, cold temperatures, too high or too low plant populations on soil with marginal N availability, insufficient and excess soil water content etc., and can confound chlorophyll meter data, thus limiting its applicability as a N test (Piekielek and Fox 1992).

Diversified Crop Rotations

Introducing legume crops in well established crop rotations has been shown to decrease N buildup in soils as well as N losses. In Iowa, Baker and Melvin (1994) documented much lower nitrate-N concentrations beneath alfalfa than for corn or soybean. Similarly, changing continuous corn to a corn-soybean rotation has been shown to reduce N losses in field (Randall et al. 1997).

Factors contributing to reduced potential for nitrate accumulation and soil N loss for various crop rotations extend beyond fertilizer practices. Interactions between hydrology and tillage are very important because any residual nitrate that accumulates in soil profile, whether from N fertilizer or microbial processes, can be lost if it is not assimilated by microbes decomposing crop residue or taken up by another plant. The rate of N cycling is important because although N-fixing legumes can release large quantities of N to soils over time, organic N derived from plant and microbial residues is not as rapidly available to plants as inorganic

N provided by most commercial fertilizers. Additionally, the gradual release of organic N is often better synchronized with subsequent plant needs and microbial population dynamics than point-in-time applications of N fertilizers. The large flush of available N following an inorganic fertilizer N application can often supply more N than can be assimilated by plants and microbes. When this pool is nitrified, large amounts of nitrate are susceptible to accumulation and losses, and can potentially contaminate surface and groundwater resources. These processes are not limited to row crop areas because research also suggests that if excessive N is applied to either annual or perennial forage crops, crop and water quality can be degraded by a resultant increase in soil nitrate concentrations (Anderson et al. 1997).

Cover Crops

Cover crops have been shown to reduce the potential for N losses from farm fields (Magdoff 1991; Staver and Brinsfield 1998) by mimicking natural ecosystems such as prairies where some plant species are growing whenever the ground is not frozen. They function by accumulating the inorganic soil N between main crop seasons and holding it in an organic form, thus preventing it from loss or build up in soil. The N is subsequently released to the next crop as the cover crop residue decomposes. Cover crops also protect against soil erosion (Dabney 1998; Kaspar et al. 2001), increase SOM (Reicosky and Forcella 1998; Dinesh et al. 2004), and suppress weed growth (Buhler et al. 1998; Lal et al. 1991). Desirable attributes for cover crops include the ability to establish rapidly under less-than-ideal conditions like cool temperatures and decreasing day length and not inhibit the growth of subsequent row crops.

In studies reviewed by Meisinger et al. (1991), cover crops reduced both the mass of N leached and nitrate concentration of leachate by 20 to 80% compared with no cover crop control. They also determined that grasses and brassicas were two to three times more effective than legumes in reducing the potential of soil N build up and N losses. Rye (*Secale cereale* L.) has also been used successfully as a cover crop in the corn and soybean belt in the US. But since rye overwinters, it must be killed, otherwise it can reduce the yield of subsequent corn crops by using too much water in the spring or immobilizing soil N (Munawar et al. 1990; Karlen and Doran 1991; Johnson et al. 1998). Other evidence suggests that the corn yield depression caused by the rye cover crop may be due to allelochemicals (Tollenaar et al. 1993). Oat has also been demonstrated to be an effective cover crop because its seed is easy to obtain and is inexpensive, fall growth is more vigorous than rye, and it winter kills, thus eliminating the need for herbicide or tillage in spring (Johnson et al. 1998).

Other plant species, including legumes, cereals, grasses, and brassicas have been evaluated and used as cover crops, but these species have generally not been successful as true cover crops in the northern corn and soybean belt. One exception was a study done in Wisconsin where hairy vetch (*Vicia villosa* Roth) and red clover grown in an oat-corn rotation showed a favorable economic comparison with continuous corn fertilized with 180 kg N/ha commercial N fertilizer (Stute and Posner 1995).

Plant Residue Management

Management of plant residues is directly related to tillage and affects N cycling in soils because plant tissue is a primary source and sink for C and N. Normally, when plant residues with C/N ratios greater than approximately 20:1 are added to the soil, available N is immobilized during the first few weeks of decomposition (Sinha et al. 1977; Doran and Smith 1991; Somda et al. 1991; Green and Blackmer 1995). Green et al. (1995) observed that incorporation of corn stover into soil resulted in rapid immobilization of all available inorganic N during the decomposition period. This occurred because the microbial population decomposing the plant residue had increased its biomass in response to the C source. The availability of nitrates for the plants may be reduced if such immobilization occurs when plants need N for growth and development. When this involves the primary crop, the immobilization process can have a negative economic impact by ultimately limiting the yield. Eventually, as residue decomposition proceeds, the C/N ratio will begin to approach that of SOM (~10:1 or 12:1), microbial biomass will decrease, and N from plant residues that was incorporated in the microbial biomass will once again be released into the soil. However, if temperature, water content, or other factors slow the residue decomposition process, N may not be released from the plant residue or microbial biomass until the primary crop has matured and stopped assimilating N. This was hypothesized by Varvel and Peterson (1990) as a factor affecting continuous corn production when they estimated that 80% of the applied fertilizer N was still immobilized in crop residues, SOM, and microbial biomass at the end of the growing season. This emphasizes the importance of understanding all of the factors affecting plant residue decomposition and how they might be manipulated to reduce soil nitrate build-up and loss without decreasing the availability of N to the primary crop or adversely affecting the soil C and N pools.

Disturbance of plant residue through tillage can significantly increase the decomposition rates (Douglas et al. 1980; Holland and Coleman 1987; Aulakh et al. 1991). Additionally, root residues may respond differently to tillage or disturbance than shoot residues. For example, Martin (1989) observed that decomposition of root residues was more rapid and more complete when they were left undisturbed in the soil than when air-dried roots were mixed with moist or air-dried soil.

Even in the absence of disturbance, decomposition rates of root and shoot residue may differ. In a laboratory simulated no-till experiment, Gale and Cambardella (2000) found that 75% of the new C inputs into soil after 1 yr of decomposition were root derived and 25% were shoot derived. They concluded that accrual of soil organic C associated with no-till is primarily due to the greater retention of root-derived C in the soil.

Tillage Impact on Nitrogen Cycling

Although tillage affects N cycling within the soils, farmers do not intentionally use tillage to manage N. In more arid regions, where fallow is traditionally used every second or third year, increased N availability following the fallow period was often more important than the perceived water conservation associated with that soil management practice (Haas et al. 1974). Tillage alters the soil environment by aerating the zone of disturbance and increasing the availability of oxygen to soil microorganisms. This favors different microbial species, populations, and processes more than a non-tilled soil (Doran 1987). The net result of tillage is increased aerobic microbial activity, leading to elevated oxidation of SOM and mineralization of soil N (Randall et al. 1997). This N mineralization response, often associated with pre-plant tillage, is also a benefit associated with using cultivation for weed control during the growing season.

Depending on tillage to release N for crop production is generally not a wise soil management practice. From a soil quality perspective, it reduces the benefits of SOM such as cation exchange capacity, soil structure, and water retention capacity merely for the release of plant available N. It also exposes the soil surface to wind and water forces, thus increasing the potential for increased erosion (Reicosky et al. 1995). Furthermore, depending on seasonal weather patterns, temperature, and rainfall, tillage during autumn or early spring can cause N mineralization too early and increase the potential for nitrate build-up and loss before subsequent crops have an opportunity to assimilate the N provided by these processes.

Effects of tillage on N management have been demonstrated in studies comparing no-till with conventional tillage. In a long-term Minnesota study (Randall and Iragavarapu 1995), residual soil nitrate contents in the 0 to 1.5 m soil profile were significantly higher with conventional tillage than no-till for 5 out of 11 yr and were not significantly different for the other 6 yr. Average flow-weighted nitrate concentrations were 13.4 and 12.0 mg/L for conventional and no-till corn production treatments, respectively. Furthermore, while the no-till treatment had 12% greater subsurface drainage flow than the conventional treatment, nitrate losses were marginally greater (about 5%) with conventional tillage. Although insignificant, these results suggest a minimal tread toward greater nitrate losses with conventional tillage in this study. The authors concluded that nitrate losses through tile drainage depend more on growing-season precipitation than on tillage. More recently, Randall and Mulla (2001) concluded that nitrate losses from agricultural fields are minimally affected by the differences in tillage systems compared with N management practices.

In Iowa, Kanwar et al. (1993) monitored differences in nitrate concentration in drainage water in continuous corn and corn-soybean rotations managed using moldboard plowing, chisel plowing, ridge plowing, and no-tillage practices. The 3 yr average nitrate concentration in drainage water from continuous corn plots receiving moldboard tillage was significantly greater (35.8 mg/L) than for either no-till or ridge tillage treatments (22.2 and 21.9 mg/L, respectively). Similar trends were observed for the corn-soybean rotation where the 3-yr average nitrate concentrations in drainage water were 19.0, 13.4, and 13.9 mg/L for moldboard plow, no-tillage, and ridge tillage treatments, respectively. However, due to a higher volume of water moving through the soil with no-till and chisel plow, the 3 yr average nitrate N concentrations in drainage water were greater with no-till and chisel plow (61.2 and 64.3 kg/ha, respectively) than with moldboard plow (45.8 kg/ha) under continuous corn. Nitrate N load losses by tillage method for the corn-soybean rotation produced trends that differed from those of continuous corn. Under the corn-soybean rotation, greater nitrate concentrations in drainage water were observed with chisel plow (32.1 kg/ha), followed by moldboard plow (27.5 kg/ha), no-till (23.9 kg/ha), and ridge tillage (23.7 kg/ha). These results led Weed and Kanwar (1996) to conclude that despite having an influence on N mineralization, tillage had less impact than crop rotation.

Nitrate Removal Strategies

If agricultural management of N by all in-field means (e.g., crop rotations, cover crops, fertilizer application best management practices, and tillage) cannot satisfactorily reduce the potential for soil nitrate build-up and loss, alternative strategies may be needed to remove nitrate from subsurface drainage, shallow groundwater, and/or surface water. Numerous methods for removing nitrate from water have been identified. These include ion exchange, biological denitrification and assimilation, chemical denitrification, reverse osmosis, electrodialysis, and catalytic denitrification (Kapoor and Viraraghavan 1997). Of these, only biological denitrification and assimilation seem suitable and adaptable for removing nitrate from subsurface drainage in agricultural fields and watersheds.

Riparian buffers

Multiple basic strategies that rely on biological denitrification and/or assimilation as either primary or secondary mechanisms for reducing nitrate concentrations in shallow subsurface waters have been and continue to be researched. One such strategy is the use of riparian buffers. Research on the effects of riparian buffers on shallow groundwater quality has been conducted in many areas across the USA and overseas. A consistent conclusion of research projects on riparian buffers is that the buffers decrease the nitrate concentration of the shallow groundwater flowing through them, in many cases to a dramatic degree, ranging from 48 to 100% (Peterjohn and Correll 1983; Jacobs and Gilliam 1985; Haycock and Pinay 1993; Hubbard and Lowrance 1997; Snyder et al. 1998; Spruill 2000). However, there are differences among these studies as to the reported primary mechanisms responsible for these reductions.

One potential factor contributing to reduced nitrate concentrations in shallow groundwater is dilution. Precipitation that infiltrates in the buffer area can contain low nitrate concentrations compared with the shallow groundwater originating from the adjacent cropland. Most studies have reported insignificant effects of dilution on the observed nitrate concentration reductions, but there have been a couple of exceptions like in the study of Hubbard and Lowrance (1997), where they estimated that 30 to 35% of the reduction was due to dilution effects.

Vegetative assimilation is one potential biological fate of nitrate in shallow groundwater as it passes through a riparian buffer. Forest trees and grass strips, which are frequently placed between the trees and cropland, do assimilate nitrate from the shallow groundwater. It is the degree to which each vegetative group assimilates nitrates that varies among the research studies. Lowrance (1992) measured a seven- to ninefold decrease in the shallow groundwater nitrate and nitrate/Cl ratio within the first 10 m of riparian forest buffer. He attributed this change in nitrate concentration primarily to forest vegetation assimilation because the denitrification potentials of the soils at the typical depths of the water table were too limited to account for the dramatic losses of nitrate.

Riparian buffers adjacent to cropland and placed in wet soils can provide several of the environmental factors that are essential to drive the processes of denitrification. Riparian buffers can supply the C required by the denitrifying bacteria, helping to fuel a biologically active zone within the soil profile. Denitrification rates have been found to be much greater within the riparian buffers placed in soils with high water tables and organic C than in well-drained soils that are more aerobic (Jacobs and Gilliam 1985; Ambus and Lowrance 1991; Gold and Groffman 1995).

An additional factor that influences denitrification is the residence time of the shallow groundwater as it passes through the soil. Areas where the flow of shallow groundwater is relatively slow are most effective for removing nitrate because of increased residence time in anaerobic and biologically active zones (Lowrance et al. 2000).

Although riparian buffers have repeatedly been proven very effective at removing nitrate from shallow groundwater, any field tile drainage line that passes through a buffer and empties directly into a surface water stream will bypass all nitrate remediation benefits of a riparian buffer. In order to utilize riparian buffer technology in tile-drained areas, alterations would be required. One option would be to terminate tile lines before entering a buffer zone. The tile flow would then seep through the buffer to the stream. Though effective, this strategy would severely limit the drainage capacity of the tile. A more appropriate option may be to discharge the tile effluent to a wetland retention area adjacent to the riparian buffer to treat the tile drainage before it enters the stream.

Wetlands

Natural and constructed wetlands have been used successfully as biological treatment systems for nitrate removal (Gersberg et al. 1983; Romero et al. 1999). Eriksson and Weisner (1997) confirmed that epiphytic biofilms on submerged vegetation could remove nitrate by denitrification. In a laboratory wetland microcosm experiment, Ingersoll and Baker (1998) determined nitrate removal efficiencies at two temperatures (28 and 35 °C), varying hydraulic-loading rates (5–20 cm/d) and C additions (1–6 g/wk dried plant residue) with nitrate contaminated water (30 mg/L nitrate-N). They measured nitrate removal efficiencies from 8% to >95%, which decreased with increasing hydraulic-loading rates and increased with increasing C addition rates. Nitrate removal efficiencies increased as the C/N ratio increased to 5:1, and at C/N ratios >5:1, nitrate loss was nearly complete. One conclusion was that the denitrification rate constant directly depended on the C addition rate. Therefore, the authors suggested that a wetland could become more efficient in nitrate removal if its plant growth were to be increased or if the plants were cut and the residue was left in place.

Crumpton et al. (1995) studied the efficacy of wetlands for removing nitrate by assimilation and denitrification. They estimated that nitrate draining from approximately 100 ha of land producing corn could potentially be removed by a 1 ha wetland. However, given an approximate 100:1 ratio of cropland to wetland area suggested by them, such a wetland may not provide sufficient residence time to remove an appreciable amount of nitrate from drainage water during high rainfall events. Xue et al. (1999) came to similar conclusions in their study of the denitrification capacity of three wetlands in Illinois. Mean monthly nitrate loading rates to the wetlands were highly variable (4–451 kg N) due to a wide range in the amounts of monthly precipitation. The ratios of denitrification capacity and mean nitrate load ranged from 19 to 59% for the three wetlands, with an average of 33%. The authors estimated that for months of low inputs, almost all of the nitrate could be denitrified, but for months of high inputs, only a small percentage would be removed.

Bioreactors

Substantial research has been done on designing bioreactors for denitrification (McCleaf and Schroeder 1995; Reising and Schroder 1996; Shanableh et al. 1997). Most designs require a supplemental C source such as sucrose (Sison et al. 1995), ethanol or acetic acid (Constantin and Fick 1997), or methane (Thalasso et al. 1997) to be effective. They also require a high level of management for in-field or edge-of-field treatment of subsurface drainage water. Solid C sources have also been tested and would appear to be more amenable to field application. Interestingly, Volokita et al. (1996) used shredded newspaper as a C source in laboratory columns to obtain N removal rates ranging from 0.056 to 0.875 mg N/g newspaper/d. Blowes et al. (1994) used a fixed-bed bioreactor filled with a compost mixture of sand, tree bard, wood chips, and leaves to treat drainage water from a farm field. Over a year, a 200-L bioreactor was able to remove nearly entire nitrate from a 10 to 60 L/d discharge of field drainage water containing 3 to 6 mg/L nitrate-N. Neither study, however, differentiated between assimilation and denitrification as nitrate loss mechanism.

Robertson and Cherry (1995) demonstrated the nitrate removal potential of a bioreactor constructed *in situ*. Very high nitrate-N concentrations (57–62 mg/L) were reduced to much lower concentrations (2–25 mg/L) in groundwater passing laterally through the bioreactor. They attributed the nitrate removal to heterotrophic denitrification, with the sawdust serving as a labile C source, and estimated that this denitrification wall would have an effective lifetime of 20 to 200 yr. However, they offered only indirect evidence that denitrification was the primary removal mechanism, citing only reduced oxygen and sulfate concentrations in water passing through the wall. In a similar study with a constructed denitrification wall, Schipper and Vojvodic-Vukovic (1998) found nitrate-N concentrations to be reduced from between 5 and 16 mg/L to <2 mg/L in shallow groundwater passing through a wall. They attributed the nitrate removal to denitrification and reported that denitrifying enzyme activity reached a plateau of 906 ng/g/h after 6 months of operation.

FUTURE RESEARCH NEEDS

The logistics and time required for soil sampling and analysis in relation to the window of opportunity for fertilizer application has prevented widespread adoption of the PSNT-LSNT N management approach. A test is needed that can be performed earlier in the season and that relies on real-time meteorological and soil data to predict plant-available soil N status in late spring. This would allow for more timely soil sampling and would increase the window of opportunity for N application. One approach to developing a model to predict soil N mineralization from soil and meteorological data is to use soil respiration measurements as a surrogate for measurements of soil N mineralization (Parkin et al. 1996). As soil N mineralization is the direct result of microbial activity, measurements of soil respiration without actively growing plants present are an integrated measure of soil microbial activity and are likely correlated with soil N mineralization. The advantage of using soil respiration measures as a surrogate for soil N mineralization measurements is that soil respiration can be measured almost continuously, responds quickly to changes in meteorological conditions, and can be used to determine cumulative activity of microorganisms over very short or very long time periods. This approach would be more amenable to determining the effect of meteorological conditions on soil N mineralization than direct measurements of changes in soil nitrate concentrations, which must be taken at discrete points in time. Coupling respiration measurements with meteorological data and estimates of soil water movement should allow the prediction of soil nitrate accumulation.

A model that predicts N mineralization based on temperature (growing degree days) has been developed (Honeycutt et al. 1988). However, application of the model requires a calibration procedure for the soil of interest. This calibration involves lengthy incubations of soil samples (30 to 60 d) to determine N mineralization potential, and thus precludes widespread adoption and use. However, N mineralization potential is often not fully realized due to variations in climatic conditions. Soil respiration measurements may be a means of *in situ* calibration. Improving our understanding of N immobilization and nitrification processes in soils and N assimilation processes in plants would also facilitate the development of predictive models based on interactions between local soils and weather patterns, and thus reduce or eliminate the need for PSNT-LSNT sampling and soil analysis.

Pierce and Nowak (1999) stated that there appears to be no standards regarding the underlying agronomic principles that should be guiding the development and application of precision agriculture and variable rate N application. At this time also, there appears to have been little advancement in regard to this problem, which indicates that much research is needed in this area if precision farming techniques are adopted and successfully practiced by farmers (Johnston et al. 2009). Improvements in long-range weather prediction models and incorporation of soil drainage and organic matter decomposition for various soil types are needed to create more reliable guidelines for variable-rate N management. Geographic information systems (GIS) and remote sensing technologies that replace chlorophyll meters may play an integral role in establishing variable-rate strategies, but further research is needed to refine the technologies and their application for agricultural purposes. To be effective, remote sensing must be able to correctly differentiate crop areas that are N deficient due to low soil N from conditions, such as water-saturated soils, Mg and K deficiencies, or disease and insect infestations, which may lead to chlorosis in plants. Currently, it remains to be seen whether sensing across multiple wavelengths (hyperspectral remote sensing) can separate these factors and correctly identify those areas where soils are N deficient.

Cultural and plant-breeding research is needed to identify and improve the compatibility of cover crops with current cropping systems. For example, rye cover crops grow well in the northern corn and soybean belt in the US, but rye genotypes that do not reduce corn yields need to be found. A better understanding of the timing of cover crop N release from shoot and root residues as well as their interactions with tillage and surface conditions is needed. Agronomic research focusing on improved establishment of cover crops, reduced management costs, and decreased negative impacts on subsequent row crops is also needed. Also, research will be needed to address a variety of limitations for wetlands, buffers, and bioreactors, including residence time for remediation and assimilation, available C sources, land area limitations, and costs of implementation. Also, because the predominant types of limitations will differ geographically, a suite of options will need to be developed rather than a single solution.

SUMMARY AND CONCLUSION

The most desirable outcome associated with research and technology development to reduce nitrate accumulation in plants is a better public recognition that several different management practices will be required for this purpose. The problem is not solely caused by N fertilizer management or by any other single factor, but is a combination of soil management practices and inherent physical, chemical, and biological characteristics of the soil as well as the physiological and environmental factors. Several current and potential N management practices are being evaluated and used to reduce the nitrate loading. Based on current experiences, a combination of two or more management practices may work in harmony and also may make implementation of the practices more cost effective.

REFERENCES

Abbasi, M.K., Shah, Z. and Adams, W.A. (2003). Effect of the nitrification inhibitor nitrapyrin on the fate of nitrogen applied to a soil incubated under laboratory conditions. *J. Plant Nutr. Soil Sci.* **166(4):** 513-518.

Ambus, P. and Lowrance, R. (1991). Comparison of denitrification in two riparian soils. *Soil Sci. Soc. Am. J.* **55:** 994-997.

Anderson, I.C., Buxton, D.R., Karlen, D.L. and Cambardella, C.A. (1997). The effects of cropping system on nitrogen removal, soil nitrogen, aggregate stability, and subsequent corn yield. *Agron. J.* **89:** 881-886.

Angus, J.F., St. Groth, C.F.D. and Tasic, R.C. (1990). Between-farm variability in yield responses to inputs of fertilizers and herbicide applied to rainfed lowland rice in the Philippines. *Agric. Ecosyst. Environ.* **30:** 219-234.

Anjana, Umar, S. and Iqbal, M. (2007). Nitrate accumulation in plants, factors affecting the process, and human health implications. *A Review Agron. Sustain. Dev.* **27:** 45-57.

Aulakh, M.S., Doran, J.W., Walters, D.T., Mosier, A.R. and Francis, D.D. (1991). Crop residue type and placement effects on denitrification and mineralization. *Soil Sci. Soc. Am. J.* **55:** 1020-1025.

Baker, J.L. and Melvin, S.W. (1994). Chemical management, status, and findings. *In:* Agric drainage well research and demonstration project–annual report and project summary, Iowa Dept. of Agric. and Land Stewardship, Des Moines, IA, USA, pp. 27-60.

Baker, J.L., Colvin, T.S., Marley, S.J. and Dawelbeit, M. (1985). Improved fertilizer management with a point-injector applicator. *In:* Proc 1985 Winter Meetings of *Am. Soc. Agric. Eng.*, Chicago, IL, 17-20 Dec 1985, *Am. Soc. Agric. Eng.*, St. Joseph, MI, USA.

Bijay-Singh, Yadvinder-Singh, Ladha, J.K., Bronson, K.F., Balasubramanian, V., Jagdeep-Singh, and Khind, C.S. (2002). Chlorophyll meter and leaf color chart-based nitrogen management for rice and wheat in northwestern India. *Agron. J.* **94:** 821-829.

Bijeriego, M., Hauck, R.D. and Olson, R.A. (1979). Uptake, translocation and utilization of ^{15}N-depleted fertilizer in irrigated corn. *Soil Sci. Soc. Am. J.* **43:** 528-533.

Blackmer, A.M. and Mallarino, A.P. (1997). Cornstalk testing to evaluate nitrogen management. PM–1584, Iowa State Coop Ext, Iowa State Univ, Ames, Iowa, USA.

Blackmer, A.M. and Sanchez, C.A. (1988). Response of corn to nitrogen-15-labeled anhydrous ammonia with and without nitrapyrin in Iowa. *Agron. J.* **80:** 95-102.

Blackmer, A.M., Voss, R.D. and Mallarino, A.P. (1997). Nitrogen fertilizer recommendations for corn in Iowa. PM–1714, Iowa State Coop Ext, Iowa State Univ, Ames, Iowa, USA.

Blowes, D.W., Robertson, W.D., Ptacek, C.J. and Merkley, C. (1994). Removal of agricultural nitrate from tile-drainage effluent water using in-line bioreactors. *J. Contam. Hydrol.* **15:** 207-221.

Boyd, P.M., Hanna, H.M., Baker, J.L., White, M. and Colvin, T.S. (2000). Field evaluation of anhydrous ammonia distribution manifolds. ASAE Paper 001140. 2000 ASAE Int Meet, Milwaukee, WI, *Am. Soc. Agric. Eng.*, St. Joseph, MI, USA.

Buhler, D.D., Kohler, K. and Foster, M.S. (1998). Spring-seeded smother plants for weed control in corn and soybean. *J. Soil Water Conserv.* **53:** 272-275.

Bundy, L.G., Walters, D.T. and Olness, A.E. (1999). Evaluation of soil nitrate tests for predicting corn nitrogen response in the north central region. North Central Regional Res Publ 342, Wisconsin *Agric. Exp. Stn.,* Madison, WI, USA.

Cassman, K.G., Dobermann, A., Sta. Cruz, P.C., Gines, H.C., Samson, M.I. and Descalsota, J.P. et al. (1996a). Soil organic matter and indigenous nitrogen supply of intensive irrigated rice systems in the tropics. *Plant Soil* **182:** 267-278.

Cassman, K.G., Gines, H.C., Dizon, M., Samson, M.I. and Alcantara, J.M. (1996b). Nitrogen-use efficiency in tropical lowland rice systems: Contributions from indigenous and applied nitrogen. *Field Crops Res.* **47:** 1-12.

Christensen, R.H. and Huffman, J.R. (1992). Response of corn to preplant application of nitrogen and nitrogen plus nitrapyrin. *J. Prod. Agric.* **5:** 352-358.

Constantin, H. and Fick, M. (1997). Influence of C-source on the denitrification rate of a high-nitrate concentrated industrial wastewater. *Water Res.* **31:** 583-589.

Crumpton, W.G., Baker, J.L., Owens, J., Rose, C. and Stenback, J. (1995). Wetlands and streams as off-site sinks for agricultural chemicals. *In:* Clean water-clean environment-21st century, Vol. 1, Proc Conf Working Group on Water Quality, USDA, Kansas City, MO, 5-8 March 1995, *Am. Soc. Agric. Eng.*, St. Joseph, MI, USA, pp. 49-52.

Dabney, S.M. (1998). Cover crop impacts on watershed hydrology. *J. Soil Water Conserv.* **53:** 207-213.

Dapoigny, L., Zelarney, P.T. and Hall, A.H. (2000). Effect of nitrogen nutrition on growth and nitrate accumulation in lettuce (*Lactuca sativa* L.) under various conditions of radiation and temperature. *Agronomie* **20:** 843-855.

Dawe, D. and Moya, P.F. (1999). Variability of optimal nitrogen applications for rice. *In:* Program report for 1998. IRRI, Makati City, the Philippines, pp. 15-17.

Dinesh, R., Suryanarayana, M.A., Chaudhuri, S.G. and Sheeja, T.E. (2004). Long-term influence of leguminous cover crops on the biochemical properties of a sandy clay loam Fluventic Sulfaquent in a humid tropical region of India. *Soil Tillage Res.* **77(1):** 69-77.

Dinnes, D.L., Jaynes, D.B., Colvin, T.S. and Meek, D.W. (1998). Potential for nitrogen fertilizer variable rate management: A case study. Integrated Crop Manage, April 1998, pp. 11-12.

Dinnes, D.L., Karlen, D.L., Jaynes, D.B., Kaspar, T.C., Hatfield, J.L., Colvin, T.S. and Cambardelaa, C.A. (2002). Nitrogen management strategies to reduce nitrate leaching in tile-drained Midwestern soils. *Agron. J.* **94:** 153-171.

Dobermann, A. (2000). Future intensification of irrigated rice systems. *In:* Sheehy, J.E. et al. (Ed.). Redesigning rice photosynthesis to increase yield. IRRI/Elsevier, Makati City, the Philippines/Amsterdam, pp. 229-247.

Dobermann, A., Witt, C., Dawe, D., Abdulrachman, S., Gines, H.C. and Nagarajan, R. et al. (2002). Site-specific nutrient management for intensive rice cropping systems in Asia. *Field Crops Res.* **74:** 37-66.

Doran, J.W. (1987). Microbial biomass and mineralizable nitrogen distributions in no-tillage and plowed soils. *Biol. Fertil. Soils* **5:** 68-78.

Doran, J.W. and Smith, M.S. (1991). Role of cover crops in nitrogen cycling. In: Hargrove WL (ed) Cover crops for clean water. Proc. Int. Conf., Jackson, TN, 9-11 April 1991, Soil Water Conserv Soc, Ankeny, IA, USA, pp. 85-90.

Douglas, J.T., Gross, M.J. and Hill, D. (1980). Measurement of pore characteristics in a clay soil under ploughing and direct tilling, including use of a radioactive tracer (144-Ce) technique. *Soil Tillage Res.* **1:** 11-18.

Durieux, R.P., Brown, H.J., Stewart, E.J., Zhao, J.Q., Jokela, W.E., Magdoff, F.R. (1995). Implications of nitrogen management strategies for nitrate leaching potential: Roles of nitrogen source and fertilizer recommendation system. *Agron. J.* **87:** 884-887.

Eriksson, P.G. and Weisner, S.E.B. (1997). Nitrogen removal in a wastewater reservoir: Importance of denitrification by epiphytic biofilms on submerged vegetation. *J. Environ. Qual.* **26:** 905-910.

Fee, R. (1997). Outdated methods sabotage nitrogen application. Successful Farming, Sept 1997, pp. 50-53.

Ferguson, R.B., Cahoon, J.E., Hergert, G.W., Peterson, T.A., Gotway, C.A. and Hartford, A.H. (1995). Managing spatial variability with furrow irrigation to increase nitrogen use efficiency. *In:* Robert, P.C. et al. (Ed.). Site-specific management for agricultural systems. ASA-CSSA-SSSA, Madison, WI, USA, pp. 443-464.

Fox, R.H., Roth, G.W., Iverson, K.V. and Piekielek, W.P. (1989). Comparison of soil and tissue nitrate tests for predicting soil nitrogen availability of corn. *Agron. J.* **81:** 971-974.

Gale, W.J. and Cambardella, C.A. (2000). Carbon dynamics of surface residue- and root-derived organic matter under simulated no-till. *Soil Sci. Soc. Am. J.* **64:** 190-195.

Gersberg, R.M., Elkins, B.V. and Goldman, C.R. (1983). Nitrogen removal in artificial wetlands. *Water Res.* **17:** 1009-1014.

Gold, A.J. and Groffman, P.M. (1995). Groundwater nitrate removal in riparian buffer zones. *In:* Clean water-clean environment-21st century, Vol. 1, Proc Conf Working Group on Water Quality, USDA, Kansas City, MO, 5-8 March 1995, *Am. Soc. Agric. Eng.*, St. Joseph, MI, USA, pp. 63-66.

Green, C.J. and Blackmer, A.M. (1995). Residue decomposition effects on nitrogen availability to corn following corn or soybean. *Soil Sci. Soc. Am. J.* **59:** 1065-1070.

Green, C.J., Blackmer, A.M. and Horton, R. (1995). Nitrogen effects on conservation of carbon during corn residue decomposition in soil. *Soil Sci. Soc. Am. J.* **59:** 453-459.

Gupta, S.K. (2006). Nitrate toxicity and human health. *In:* Proc Workshop on Nitrogen in Environment, Industry, and Agriculture, New Delhi, India, pp. 8-10.

Gupta, S.K., Gupta, R.C., Gupta, A.B., Seth, A.K., Bassin, J.K. and Gupta, A. (2000). Recurrent acute respiratory tract infection in areas having high nitrate concentration in drinking water. *Environ. Health Persp.* **108:** 363-366.

Haas, H.J., Willis, W.O. and Bond, J.J. (1974). General relationships and conclusions. In: Summer fallow in the western United States. Conserv Res Rep 17, USDA-ARS, Washington DC, USA, pp. 149-160.

Hancock, D.W. (2007). Nitrate toxicity. University of Georgia Cooperative Extension Publ., USA pp. 1-4 [http://74.125.153.132/search?q=cache:nvwguqh-3ogJ:pubs.caes.uga.edu/caespubs/pubs/PDF/C915.pdf+hancock+nitrate+toxicity&cd=2&hl=en&ct=clnk&gl=in&client=firefox-a].

Haycock, N.E. and Pinay, G. (1993). Groundwater nitrate dynamics in grass and poplar vegetated riparian buffer strips during the winter. *J. Environ. Qual.* **22:** 273-278.

Hendrickson, L.L., Walsh, L.M. and Keeney, D.R. (1978). Effectiveness of nitrapyrin in controlling nitrification in fall and spring applied anhydrous ammonia. *Agron. J.* **70:** 704-708.

Holland, E.A. and Coleman, D.C. (1987). Litter placement effects on microbial and organic matter dynamics in an agroecosystem. *Ecology* **68:** 425-433.

Honeycutt, C.W., Zibilske, L.M. and Clapham, W.M. (1988). Heat units for describing carbon mineralization and predicting net nitrogen mineralization. *Soil Sci. Soc. Am. J.* **52:** 1346-1350.

Hubbard, R.K. and Lowrance, R. (1997). Assessment of forest management effects on nitrate removal by riparian buffer systems. *Trans. ASAE* **40:** 383-391.

Ingersoll, T.L. and Baker, L.A. (1998). Nitrate removal in wetland microcosms. *Water Res.* **32:** 677-684.

Iowa Department of Agriculture and Land Stewardship, (1997). Agriculture drainage well research and demonstration project. Ann Rep, Iowa Dept of Agric and Land Stewardship, Des Moines, USA.

Jacobs, T.C. and Gilliam, J.W. (1985). Riparian losses of nitrate from agricultural drainage waters. *J. Environ. Qual.* **14:** 472-478.

Johnson, T.J., Kaspar, T.C., Kohler, K.A., Corak, S.J. and Logsdon, S.D. (1998). Oat and rye overseeded into soybean as fall cover crops in the upper Midwest. *J. Soil Water Conserv.* **53:** 276-279.

Johnston, A.M., Khurana, H.S., Majumdar, K. and Satyanarayana, T. (2009). Site-Specific Nutrient Management – Concept, Current Research and Future Challenges in Indian Agriculture. *J. Indian Soc. Soil Sci.* **57(1):** 1-10.

Kanwar, R.S., Karlen, D.L., Cambardella, C.A., Colvin, T.S. and Pederson, C. (1996). Impact of manure and N-management systems on water quality. *In:* Proc Annu Integrated Crop Manage Conf, 8th Ames, IA, USA, 19-20 Nov 1996, Iowa State Univ Ext Ames, pp. 65-77.

Kanwar, R.S., Stolenberg, D.E., Pfeiffer, R., Karlen, D.L., Colvin, T.S. and Simpkins, (1993). Transport of nitrate and pesticides to shallow groundwater systems as affected by tillage and crop rotation practices. *In:* Proc. Natl. Conf. on Agric. Res. to Protect Water Quality, pp. 270-273.

Kapoor, A. and Viraraghavan, T. (1997). Nitrate removal from drinking water—review. *J. Environ. Eng.* **123:** 371-380.

Karlen, D.L. and Doran, J.W. (1991). Cover crop management effects on soybean and corn growth and nitrogen dynamics in an on-farm study. *Am. J. Altern. Agric.* **6:** 71-82.

Karlen, D.L,, Kramer, L.A. and Logsdon, S.D. (1998). Field-scale nitrogen balances associated with long-term continuous corn production. *Agron. J.* **90:** 644-650.

Kaspar, T.C., Radke, J.K. and Laflen, J.M. (2001). Small grain cover crops and wheel traffic effects on infiltration, runoff, and erosion. *J. Soil Water Conserv.* **56:** 160-164.

Khurana, H.S., Phillips, S.B., Bijay-Singh, Alley, M.M., Dobermann, A., Sidhu, A.S., Yadvinder-Singh, and Peng, S. (2008). Agronomic and economic evaluation of site-specific nutrient management for irrigated wheat in northwest India. *Nutr. Cycl. Agroecosyst.* **82:** 15-31.

Khurana, H.S., Phillips, S.B., Bijay-Singh, Dobermann, A., Sidhu, A.S., Yadvinder-Singh, and Peng, S. (2007). Performance of site-specific nutrient management for irrigated, transplanted rice in northwest India. *Agron. J.* **99(6):** 1436-1447.

Kidwaro, F.M. and Kephard, K.D. (1998). Retention of nitrogen from stabilized anhydrous ammonia in the soil profile during winter wheat production in Missouri. *Comm. Soil Sci. Plant Anal.* **29:** 481-499.

Kitchen, N.R., Hughes, D.F., Sudduth, K.A. and Birrell, S.J. (1995). Comparison of variable rate to single rate nitrogen fertilizer application: Corn production and residual soil

NO_3–N. *In:* Robert, P.C. et al. (Ed.). Site-specific management for agricultural systems. ASA-CSSA-SSSA, Madison, WI, USA, pp. 427-439.

Kitur, B.K., Smith, M.S., Blevins, R.L. and Frye, W.W. (1984). Fate of 15N depleted ammonium nitrate applied to no-tillage and conventional tillage corn. *Agron. J.* **76:** 240-242.

Lal, R., Regnier, E., Eckert, D.J., Edwards, W.M. and Hammond, R. (1991). Expectations of cover crops for sustainable agriculture. *In:* Hargrove, W.L. (Ed.). Cover crops for clean water. Proc Int Conf, Jackson, TN, 9-11 April 1991, Soil and Water Conserv. Soc., Ankeny, IA, USA, pp. 1-11.

Larson, W.E. and Robert, P.C. (1991). Farming by soil. *In:* Lal, R. and Pierce, F.J. (Eds.). Soil management for sustainability. Soil and Water Conserv. Soc., Ankeny, IA, USA, pp. 103-112.

Lowrance, R. (1992). Groundwater nitrate and denitrification in a coastal plain forest. *J. Environ. Qual.* **21:** 401-405.

Lowrance, R., Hubbard, R.K. and Williams, R.G. (2000). Effects of a managed three zone riparian buffer system on shallow groundwater quality in the southeastern coastal plain. *J. Soil Water Conserv.* **55:** 212-220.

Magdoff, F.R. (1991). Managing nitrogen for sustainable corn systems: Problems and Possibilities. *Am. J. Altern. Agric.* **6:** 3-8.

Magdoff, F.R., Jokela, W.E., Fox, R.H. and Griffin, G.F. (1990). A soil test for nitrogen availability in the northeastern United States. *Comm. Soil Sci. Plant Anal.* **21:** 1103-1115.

Magdoff, F.R., Ross, D. and Amadon, J. (1984). A soil test for nitrogen availability to corn. *Soil Sci. Soc. Am. J.* **48:** 1301-1304.

Majumdar, D. (2002). Suppression of nitrification and N_2O emission by karanjin—a nitrification inhibitor prepared from karanja (Pongamia glabra Vent.). *Chemosphere* **47(8):** 845-850.

Martin, J.K. (1989). *In situ* decomposition of root-derived carbon. *Soil Biol. Biochem.* **21:** 973-974.

McCleaf, P.R. and Schroeder, E.D. (1995). Denitrification using a membrane-immobilized biofilm. *J. Am. Waterworks Assoc.* March 1995, pp. 77-86.

Meisinger, J.J., Bandel, V.A., Stanford, G. and Legg, J.O. (1985). Nitrogen utilization of corn under minimal tillage and moldboard plow tillage: I. Four year results using labeled fertilizer on an Atlantic Coastal Plain soil. *Agron. J.* **77:** 602-611.

Meisinger, J.J., Hargrove, W.L., Mikkelson, R.L., Williams, J.R. and Benson, V.W. (1991). Effects of cover crops on groundwater quality. *In:* Hargrove, W.L. (Ed.). Cover crops for clean water. Proc Int Conf, Jackson, TN, 9-11 April 1991, Soil and Water Conserv. Soc., Ankeny, IA, USA, pp. 57-68.

Munawar, A., Blevins, R.L., Frye, W.W. and Saul, M.R. (1990). Tillage and cover crop management for soil conservation. *Agron. J.* **82:** 773-777.

NAAS, (2005). Policy options for efficient nitrogen use. Policy Paper No. 33, *Natl. Acad. Agric. Sci.* New Delhi, India, pp. 1-4.

Nosengo, N. (2003). Fertilized to death. *Nature* **425:** 894-895.

Oberle, S.L. and Keeney, D.R. (1990). Factors influencing corn fertilizer N requirements in the northern U.S. corn belt. *J. Prod. Agric.* **3:** 527-534.

Olson, R.V. (1980). Fate of tagged nitrogen fertilizer applied to irrigated corn. *Soil Sci. Soc. Am. J.* **55:** 1616-1621.

Pan, W.L., Huggins, D.R., Malzer, G.L., Douglas, Jr. C.L. and Smith, J.L. (1997). Field heterogeneity in soil-plant relationships: Implications for site-specific management. *In:* Pierce, F.J. and Sadler, E.J. (Eds.). The state of site-specific management for agriculture. ASA-CSSA-SSSA, Madison, WI, USA, pp. 81-99.

Parkin, T.B., Doran, J.W. and Franco-Vizcaino, E. (1996). Field and laboratory tests of soil respiration. *In:* Methods for assessing soil quality. SSSA Spec Publ. 49, SSSA, Madison, WI, USA.

Peterjohn, W.T. and Correll, D.L. (1983). Nutrient dynamics in an agricultural watershed: Observations on the role of a riparian forest. *Ecology* **65:** 1466-1475.

Piekielek, W.P. and Fox, R.H. (1992). Use of a chlorophyll meter to predict sidedress nitrogen requirements for maize. *Agron. J.* **84:** 59-65.

Pierce, F.J. and Nowak, P. (1999). Aspects of precision agriculture. *Adv. Agron.* **67:** 1-85.

Power, J.F., Flowerday, A.D., Wiese, R.A. and Watts, D.G. (1998). Agricultural nitrogen management to protect water quality. Bull 4, USDA-CREES-IDEA, Washington DC, USA.

Power, J.F., Wiese, R.A. and Flowerday, A.D. (2000). Managing nitrogen for water quality–lessons from the Management Systems Evaluation Area. *J. Environ. Qual.* **29:** 355-366.

Randall, G.W. (1997). Nitrate-N in surface waters as influenced by climatic conditions and agricultural practices. *In:* Proc. Agric. and Hypoxia in the Mississippi Watershed Conf, St Louis, MO. 14-15 July 1997, Am Farm Bureau Fed, Park Ridge, IL.

Randall, G.W., Huggins, D.R., Russelle, M.P., Fuchs, D.J., Nelson, W.W. and Anderson, J.L. (1997). Nitrate losses through subsurface tile drainage in conservation reserve program, alfalfa, and row crop systems. *J. Environ. Qual.* **26:** 1240-1247.

Randall, G.W. and Iragavarapu, T.K. (1995). Impact of long-term tillage systems for continuous corn on nitrate leaching to tile drainage. *J. Environ. Qual.* **24:** 360-366.

Randall, G.W., Malzer, G.L. and Anderson, B.W. (1992). Nitrate losses to tile drainage as affected by nitrogen fertilization of corn in a corn-soybean rotation. Field research in Soil Sci Univ of Minnesota Southern Exp Stn, Waseca, MN, USA.

Randall, G.W. and Mulla, D.J. (2001). Nitrate nitrogen in surface waters as influenced by climatic conditions and agricultural practices. *J. Environ. Qual.* **30:** 337-344.

Raun, W.R. and Johnson, G.V. (1999). Improving nitrogen use efficiency for cereal production. *Agron. J.* **91:** 357-363.

Reddy, G.B. and Reddy, K.R. (1993). Fate of nitrogen-15 enriched ammonium nitrate applied to corn. *Soil Sci. Soc. Am. J.* **57:** 111-115.

Reichenberger, L. (1994). A better aim for NH_3. *Farm J.*, February 1994, pp. 12-13.

Reicosky, D.C. and Forcella, F. (1998). Cover crop and soil quality interactions in agroecosystems. *J. Soil Water Conserv.* **53:** 224-229.

Reicosky, D.C., Kemper, W.D., Langdale, G.W., Douglas, Jr. C.L. and Rasmussen, P.E. (1995). Soil organic matter changes resulting from tillage and biomass production. *J. Soil Water Conserv.* **50:** 253-261.

Reising, A.R. and Schroeder, E.D. (1996). Denitrification incorporating microporous membranes. *J. Environ. Eng.* **122:** 599-604.

Ressler, D.E., Horton, R., Baker, J.L. and Kaspar, T.C. (1997). Testing a nitrogen fertilizer applicator designed to reduce leaching losses. *Appl. Eng. Agric.* **13:** 345-350.

Robertson, W.D. and Cherry, J.A. (1995). *In situ* denitrification of septic-system nitrate using reactive porous media barriers: Field trials. *Groundwater* **33:** 99-111.

Romero, J.A., Comin, F.A. and Garcia, C. (1999). Restored wetlands as filters to remove nitrogen. *Chemosphere* **39:** 323-332.

Sanchez, C.A. and Blackmer, A.M. (1988). Recovery of anhydrous ammonia-derived nitrogen-15 during three years of corn production in Iowa. *Agron. J.* **80:** 102-108.

Sawyer, (1994). Concepts of variable rate technology with considerations for fertilizer application. *J. Prod. Agric.* **7:** 195-201.

Schippers, J.S. and Vojvodic-Vukovic, M. (1998). Nitrate removal from groundwater using a denitrification wall amended with sawdust: Field trial. *J. Environ. Qual.* **27:** 664-668.

Schrock, M.D., Oard, D.L., Taylor, R.K., Grimm, J., Trefz, M., Love, G. and Peterson, A. (1999). Lateral distribution of anhydrous ammonia with conventional and pulse-width-modulation metering systems. ASAE Paper 99 ATEC 102, *Am. Soc. Agric. Eng.*, St. Joseph, MI, USA.

Shanableh, A., Abeysinghe, D. and Hijazi, A. (1997). Effect of cycle duration on phosphorus and nitrogen transformations in biofilters. *Water Res.* **31:** 149-153.

Siambi, M.M., Karlen, D.L. and Shibles, R.M. (1999). Chlorophyll meter assessments of corn response to nitrogen management practices. *J. Iowa Acad. Sci.* **106:** 34-39.

Sinha, M.K., Sinha, D.P. and Sinha, H. (1977). Organic matter transformations in soils: V. Kinetics of carbon and nitrogen mineralization in soils amended with different organic materials. *Plant Soil* **46:** 579-590.

Sison, N.F., Hanaki, K. and Matsuo, T. (1995). High loading denitrification by biological activated carbon process. *Water Res.* **29:** 2776-2779.

Snyder, N.J., Mostaghimi, S., Berry, D.F., Reneau, R.B., Hong, S., McClellan, P.W. and Smith, E.P. (1998). Impact of riparian forest buffers on agricultural nonpoint source pollution. *J. Am. Water Resour. Assoc.* **34:** 385-395.

Somda, Z.C., Ford, P.B. and Hargrove, W.L. (1991). Decomposition and nitrogen recycling of cover crops and crop residues. *In:* Hargrove, W.L. (Ed.). Cover crops for clean water. Proc Int Conf, Jackson, TN, 9-11 April 1991, Soil and Water Conserv. Soc., Ankeny, IA, USA, pp. 103-105.

Spruill, T.B. (2000). Statistical evaluation of the effects of riparian buffers on nitrate and ground water quality. *J. Environ. Qual.* **29:** 1523-1538.

Staver, K.W. and Brinsfield, R.B. (1998). Using winter cover crops to reduce groundwater nitrate contamination in the mid-Atlantic coastal plain. *J. Soil Water Conserv.* **53:** 230-240.

Stehouwer, R.C. and Johnson, J.W. (1990). Urea and anhydrous ammonia management for conventional tillage corn production. *J. Prod. Agric.* **3:** 507-513.

Stute, J.K. and Posner, J.L. (1995). Legume cover crops as a nitrogen source for corn in an oat-corn rotation. *J. Prod. Agric.* **8:** 385-390.

Thalasso, F., Vallecillo, A., Garcia-Encina, P. and Fdz-Polanco, F. (1997). The use of methane as a sole carbon source for wastewater denitrification. *Water Res.* **31:** 55-60.

Tollenaar, M., Mihajlovic, M. and Vyn, T.J. (1993). Corn growth following cover crops: Influence of cereal cultivar, cereal removal, and nitrogen rate. *Agron. J.* **85:** 251-255.

Touchton, J.T., Hoeft, R.G., Welch, L.F., Mulvaney, D.L., Oldham, M.G. and Zajicek, F.E. (1979). N uptake and corn yield as affected by applications of nitrapyrin with anhydrous ammonia. *Agron. J.* **71:** 238-242.

Varvel, G.E. and Peterson, T.A. (1990). Residual soil nitrogen as affected by continuous, two-year, and four-year crop rotation systems. *Agron. J.* **82:** 958-962.

Varvel, G.E., Schepers, J.S. and Francis, D.D. (1997). Ability for in-season correction of nitrogen deficiency in corn using chlorophyll meters. *Soil Sci. Soc. Am. J.* **61:** 1233-1239.

Vetch, J.A., Malzer, G.L., Robert, P.C. and Huggins, D.R. (1995). Nitrogen specific management by soil condition: Managing fertilizer nitrogen in corn. *In:* Robert, P.C. et al. (Eds.). Site-specific management for agricultural systems. ASA-CSSA-SSSA, Madison, WI, USA.

Vieira, I.S., Vasconselos, E.P. and Monteiro, A.A. (1998). Nitrate accumulation, yield and leaf quality of turnip greens in response to nitrogen fertilization. *Nutr. Cycl. Agroecosyst.* **51:** 249-258.

Volokita, M., Belkin, S., Abeliovich, A. and Soares, M.I.M. (1996). Biological denitrification of drinking water using newspaper. *Water Res.* **30:** 965-971.

Wagger, M.G., Cabrera, M.L. and Ranells, N.N. (1998). Nitrogen and carbon cycling in relation to cover crop residue quality. *J. Soil Water Conserv.* **53:** 214-218.

Walker, R. (1990). Nitrates and N-nitroso compounds: A review of the occurrence in food and diet and the toxicological implications. *Food Addit. Contam.* **7:** 717-768.

Wang, G., Dobermann, A., Witt, C., Sun, Q. and Fu, R. (2001). Performance of site-specific nutrient management for irrigated rice in Southeast China. *Agron. J.* **93:** 869-878.

Weber, R.W., Grisso, R.D., Shapiro, C.A. and Kranz, W.L. (1993). Accuracy of anhydrous ammonia application. ASAE Paper 931548. *Am. Soc. Agric. Eng.*, St. Joseph, MI, USA.

Weed, D.A.J. and Kanwar, R.S. (1996). Nitrate and water present in and flowing from root-zone soil. *J. Environ. Qual.* **25:** 709-719.

Wood, C.W., Reeves, D.W., Duffield, R.R. and Edmisten, K.L. (1992). Field chlorophyll measurements for evaluation of corn nitrogen status. *J. Plant Nutr.* **15:** 487-500.

Wopereis, M.C.S., Donovan, C., Nebie´, B., Guindo, D. and N'Diaye, M.K. (1999). Soil fertility management in irrigated rice systems in the Sahel and Savanna regions of West Africa. Part I—Agronomic analysis. *Field Crops Res.* **61:** 125-145.

Xue, Y., Kovacic, D.A., David, M.B., Gentry, L.E., Mulvaney, R.L and Lindau, C.W. (1999). *In situ* measurements of denitrification in constructed wetlands. *J. Environ. Qual.* **28:** 263-269.

Index